The higher score

Year 6

# Mathematics

## SATs Question Workbook

Steph King
& Sarah-Anne Fernandes

RISING STARS

Orders: please contact Bookpoint Ltd, 130 Park Drive, Milton Park, Abingdon, Oxon OX14 4SE. Telephone: (44) 01235 400555. Email: primary@bookpoint.co.uk

Lines are open from 9 a.m. to 5 p.m., Monday to Saturday, with a 24-hour message answering service. Visit our website at www.risingstars-uk.com for details of the full range of Rising Stars publications.

Online support and queries email: onlinesupport@risingstars-uk.com

ISBN: 978 1 51044 273 3



This edition published in 2018 by Rising Stars, part of Hodder & Stoughton Ltd.

First published in 2015 by Rising Stars, part of Hodder & Stoughton Ltd.

Rising Stars UK is part of the Hodder Education Group

An Hachette UK Company

Carmelite House

50 Victoria Embankment

London EC4Y 0DZ

www.risingstars-uk.com

Impression number 10 9 8 7 6 5

Year 2022 2021 2020 2019



Authors: Steph King and Sarah-Anne Fernandes

Series Editor: Sarah-Anne Fernandes

Accessibility Reviewer: Vivien Kilburn

Cover design: Burville-Riley Partnership

Illustrations by Ann Paganuzzi

Typeset in India

Printed in Dubai

A catalogue record for this title is available from the British Library.

# Contents

*Welcome to Achieve Mathematics: The Higher Score – Question Workbook*

In this book you will find lots of practice and information to help you achieve the higher score in the Key Stage 2 Mathematics tests. You will look again at some of the same key knowledge that was in *Achieve Mathematics: The Expected Standard – Question Workbook*, but you will use it to tackle trickier questions and apply it in more complex ways.

# About the Key Stage 2 Mathematics National Tests

The tests will take place in the summer term in Year 6. They will be done in your school and will be marked by examiners – not by your teacher.

There are three papers to the tests:

**Paper 1: Arithmetic – 30 minutes (40 marks)**

- These questions assess confidence with a range of mathematical operations.
- Most questions are worth 1 mark. However, 2 marks will be available for long multiplication and long division questions.
- It is important to show your working – this may gain you a mark in questions worth 2 marks, even if you get the answer wrong.

**Papers 2 and 3: Reasoning – 40 minutes (35 marks) per paper**

- These questions test mathematical fluency, solving mathematical problems and mathematical reasoning.
- Most questions are worth 1 or 2 marks. However, there may be one question with 3 marks.
- There will be a mixture of question types, including multiple-choice, true/false or yes/no questions, matching questions, short responses such as completing a chart or table or drawing a shape, or longer responses where you need to explain your answer.
- In questions that have a method box it is important to show your method – this may gain you a mark, even if you get the answer wrong.

You will be allowed to use a pencil/black pen, an eraser, a ruler, an angle measurer/protractor and a mirror. You **are not allowed** to use a calculator in any of the test papers.

# Test techniques

## Before the tests

- Try to revise little and often, rather than in long sessions.
- Choose a time of day when you are not tired or hungry.
- Choose somewhere quiet so you can focus.
- Revise with a friend. You can encourage and learn from each other.
- Read the 'Top tips' throughout this book to remind you of important points in answering test questions.
- Keep track of your score using the table on the inside back cover of this book.

## During the tests

- READ THE QUESTION AND READ IT AGAIN.
- If you find a question difficult to answer, move on; you can always come back to it later.
- Always answer a multiple-choice question. If you really can't work out an answer, try to think of the most sensible response and read the question again.
- Check to see how many marks a question is worth. Have you written enough to 'earn' those marks in your answer?
- Read the question again after you have answered it. Make sure you have given the correct number of answers within a question, e.g. if there are two boxes for two missing numbers.
- If you have any time left at the end, go back to the questions you have missed.

Where to get help:

- **Pages 6–7** practise number and place value.
- **Pages 8–17** practise number – addition, subtraction, multiplication and division.
- **Pages 18–26** practise number – fractions, decimals and percentages.
- **Pages 27–30** practise ratio and proportion.
- **Pages 31–34** practise algebra.
- **Pages 35–43** practise measurement.
- **Pages 44–45** practise geometry – properties of shapes.
- **Pages 46–51** practise geometry – position and direction.
- **Pages 52–58** practise statistics.
- **Pages 59–63** provide the answers to the questions.

The pencil icon appears next to questions for which you should show your workings.

# Place value

To achieve the higher score, you need to:
★ know the **place value** of numbers up to 10,000,000 with up to three **decimal places.**

**1** Write the value of the digit **6** in these numbers.

4,264,000 60,000

505.362 0.06

0.576 0.006

(1 mark) 1

**2** Circle the **two** numbers that sum to 8,500,000

2,132,000 700,000 1,500,000 6,368,000 2,232,000

(1 mark) 2

**3** Tick (✓) the statement that is **not** true. Tick **one.**

The digit **8** in 483,723 has the value **80,000**

The digit **1** in 2,107,225 has the value **ten thousand.**

The digit **9** in 789,888 has the value **9,000**

The digit **4** in 2.354 has the value **four thousandths.**

(1 mark) 3

**4** 3 0 4 1 5

Use four of these cards to make a four-digit **multiple of 10** that is **less than** 3,000

Find two different ways to do this.

(1 mark) 4

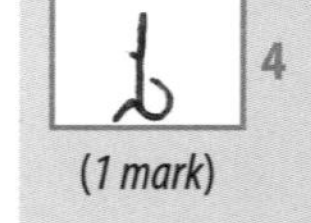

**5** Write the **third odd number** that can be placed on the shaded part of this number line. 4,960,000

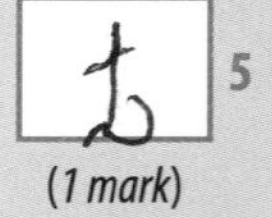

(1 mark) 5

4,900,000 5,000,000

**Top tip**

- Work out the scale on the number line first and label any points you know.

5/5 Total for this page

# Roman numerals

To achieve the higher score, you need to:
★ read **Roman numerals** to 1,000 (using I, V, X, L, C, D and M).

**1** What number do the Roman numerals CCCLXXXIII represent?

383

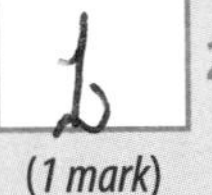

(1 mark) 1

**2** Arrange these four Roman numerals to make a Roman number that makes this number statement **true**.

    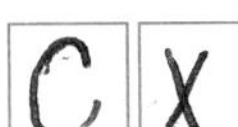 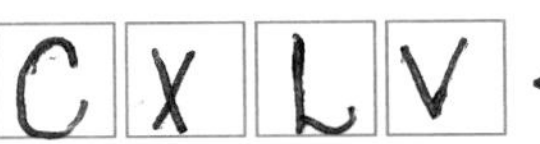 < 150

X V C L — C X L V < 150

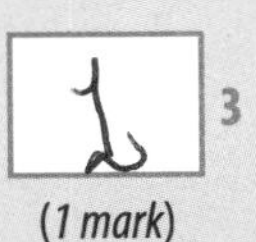

(1 mark) 2

**3** Pete writes a sequence using Roman numerals starting from 198

He uses the rule **subtract 20**

Circle all the Roman numerals that will **not** be in this sequence.

CLXXVIII  CLVII  CXLVIII CXXXVIII CXVIII

(1 mark) 3

**4** The Great Fire of London took place in 1666

a) Write the date 1666 in Roman numerals.

MDCLXVI

(1 mark) 4a

b) What do you notice about the numerals you have used?

Written in order from greatest to least
All of the 7 roman numerals are used

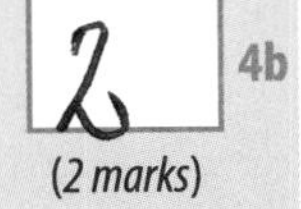

(2 marks) 4b

**5** Amber has just finished watching her favourite film with Gran.

Gran says: 'That film was made nine years after the year I was born.'

Amber thinks that Gran was born in 1957.

Is she correct? Circle your answer. YES / NO

Explain your answer.

The film was made only 9 years after Gran was born and the film was made in 1967.

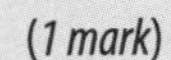

(1 mark) 5

**Top tip**

- Check the order of two numerals with different values, e.g. XC means 10 less than 100, whereas CX means 10 more than 100.

7 /7

Total for this page

# Addition and subtraction

To achieve the higher score, you need to:
★ solve multi-step addition and subtraction problems, deciding which operations and methods to use and why.

**1** Complete this sequence.

2,325 4,793 7,261 [9729] [12,197]

(1 mark) 1

**2** 23,782 + [4472] = 41,509 – 13,256

(1 mark) 2

**3** Fill in the missing digits.

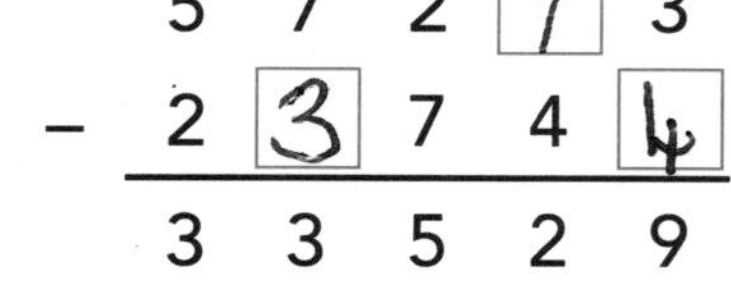

|   | 5 | 7 | 2 | [7] | 3 |
|---|---|---|---|---|---|
| – | 2 | [3] | 7 | 4 | [4] |
|   | 3 | 3 | 5 | 2 | 9 |

(1 mark) 3

**4** A small swimming pool holds 162,000 litres of water.

The pool will be filled over three days.

Complete the table.

| Day 1 | Day 2 | Day 3 |
|---|---|---|
| 59,865 litres | 59,865 litres | litres |

(1 mark) 4

**5** a) Ben has a box of **150** cubes.

He uses some of the cubes to build a tower.

**98** cubes are left over.

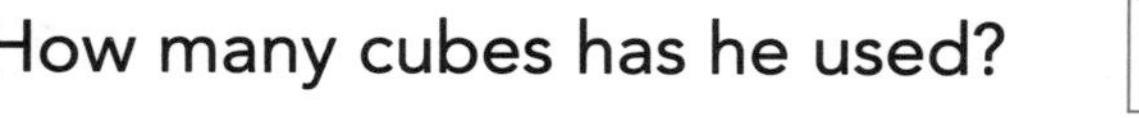

How many cubes has he used?  [52]

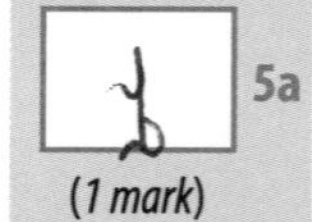

(1 mark) 5a

b) Seb has **98** cubes left over.

He builds two more towers.

One tower uses **23** cubes and the other uses **19** cubes.

How many of his **98** cubes has he got left now?  [56]

(2 marks) 5b

**Top tip**

- Remember to line up digits with the same value when using the column method.

7/7 Total for this page

# Squares and cubes

To achieve the higher score, you need to:

- ★ recognise and use **cube numbers**, and use the notation for cubed ($^3$)
- ★ solve problems involving multiplication and division using your knowledge of factors and multiples, squares and cubes.

**1** Ali thinks of a square number.

He squares it and then divides by 3

His new number is 27

What is Ali's number? 

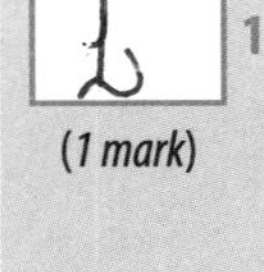

(1 mark) 1

**2** The row and column in the diagram both sum to the same square number.

Find the missing single-digit numbers to make this true.

| | 49 | |
|---|---|---|
| 32 | $8^2$ | 25 |
| | $2^3$ | |

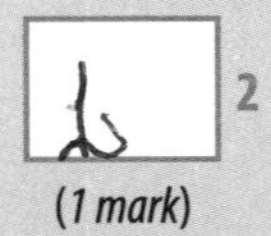

(1 mark) 2

**3** The volume of a cube is 512 cm$^3$.

What are the dimensions of the cube?

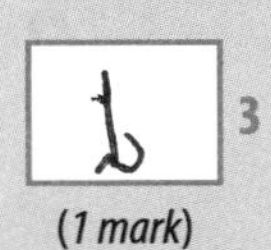

(1 mark) 3

**4** △ = 95 + ⬠

The value of the triangle is a three-digit cube number.

The value of the pentagon is a square number.

△ = 216

⬠ = 121

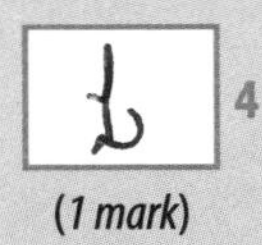

(1 mark) 4

**5** The volume of this cuboid is 360 cm$^3$.

All dimensions are in whole centimetres but are less than 10 cm.

What are the dimensions of the base?

9 cm × 5 cm

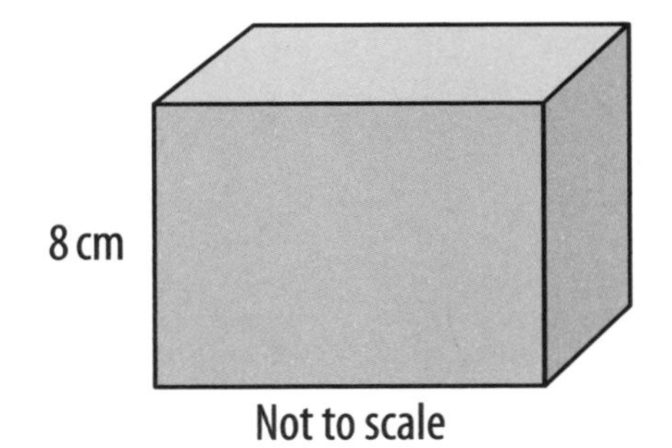


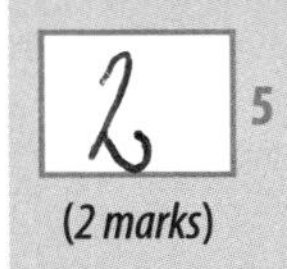

(2 marks) 5

**Top tip**

- Trial and improvement is a very useful technique for some of these questions.

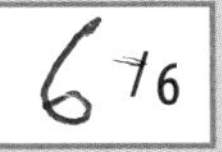

Total for this page

# Common multiples

To achieve the higher score, you need to:
★ identify **common multiples**.

**1** Nina chooses a common multiple of 8 and 12
It is **greater than 150** but **less than 200**

What number could Nina have chosen? 192

1 (1 mark)

**2** What is the **smallest** number that can be placed in the shaded circle?

72

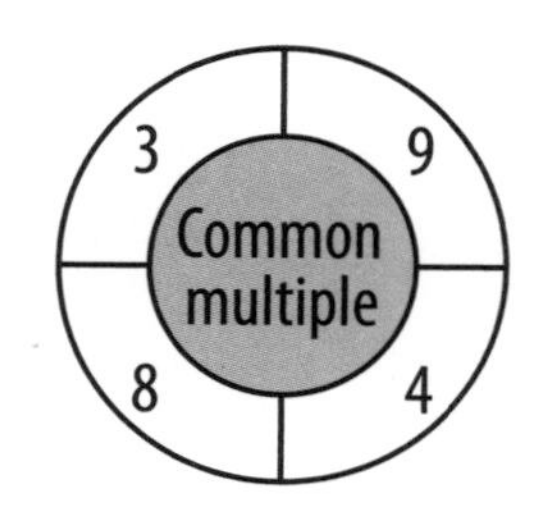


2 (1 mark)

**3** What is the first number **greater than 300** that can be placed in the shaded circle?

360

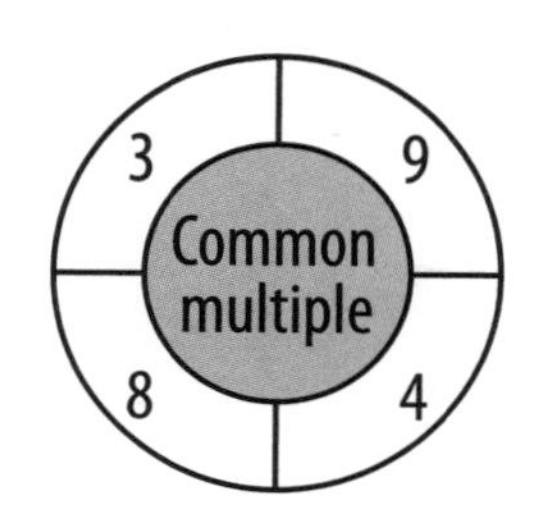


3 (1 mark)

**4** Tom can arrange his cars in groups of 9 with none left over.
He can also arrange his cars in groups of 12 with none left over.

Circle the total number of cars that Tom **cannot** have.

108 720  909 540  480

4 (1 mark)

**5**

| Length | Price |
|---|---|
| 500 mm | £3.65 |
| 840 mm | £4.95 |
| 900 mm | £5.69 |

Emily buys **two** planks of wood of the **same** length.

She cuts one into seven equal pieces. She cuts the other into six equal pieces.

All her pieces are a **multiple of 10 mm**. No wood is left over.

How much change does she receive from a £20 note? 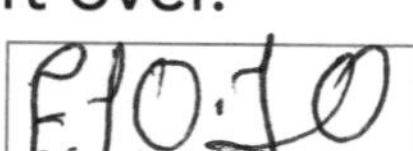

5 (3 marks)

**Top tip**

- A common multiple is found in two or more multiplication tables.

Total for this page /7

# Common factors

To achieve the higher score, you need to:

- ★ find all the **factor** pairs of a number
- ★ solve problems involving multiplication and division using your knowledge of factors.

**1** List all the factors of 48 that are also factors of 56

1, 2, 4, 8

(1 mark) 1

**2** Fill in the two missing factors so that they match the multiples shown here.

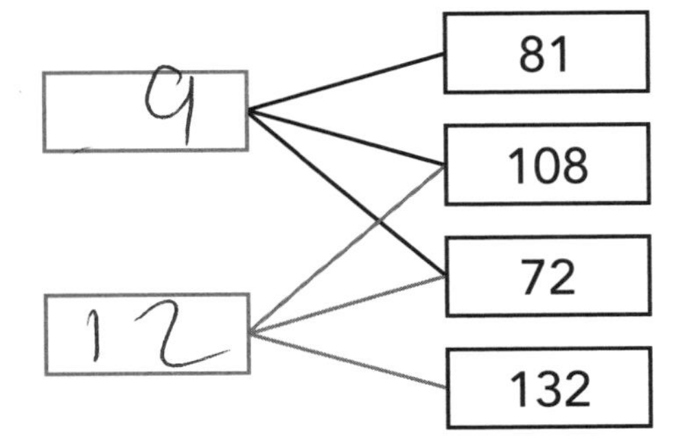


(1 mark) 1

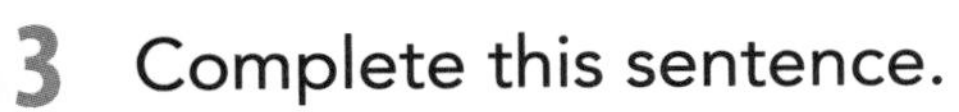

**3** Complete this sentence.

132 and 36 both share the factors 1; 2; 3; 4; 6; and 12

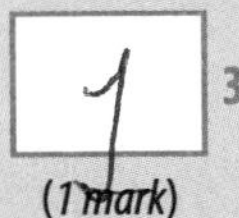

(1 mark) 1

**4** Ben arranges his 120 counters into equal groups.

Shazia also arranges her 144 counters into equal groups.

Tick (✓) the number of equal groups that both children can make.

| 4 ✓ | 10 | 8 ✓ | 6 ✓ | 9 |
|---|---|---|---|---|

(1 mark) 1

**5** Jade finds all the factors of 280

Write all the factors of 280 that are also multiples of 10

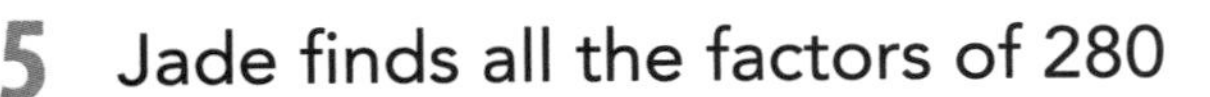

10, 20, 40, 70, 140, 280

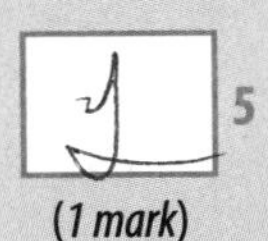

(1 mark) 1

**Top tip**

- Notice that question 5, and others like it, combine both multiples and factors.

5/5 Total for this page

# Prime numbers and prime factors

To achieve the higher score, you need to:
- ★ identify and use **prime numbers** up to 100
- ★ recognise and use **prime factors**.

**1** Circle **all** the prime numbers.

51 57 61 67 69 71

(1 mark) 1

**2** The prime factors of a number are 2; 3; 5; and 7

What is the number? ☐

(1 mark) 2

**3** The numbers in this grid are all prime factors.

The prime factors 2; 2; 3; and 5 are missing.

Complete the grid so that the **product** of each row and each column is a multiple of **10**

| 3 | 5 | ☐ | 5 |
|---|---|---|---|
| 5 | ☐ | 2 | 2 |
| 2 | ☐ | 5 | ☐ |

(1 mark) 3

**4** George writes down two square numbers.

Both numbers have the prime factors 2 and 3 in common.

What could George's numbers be? ☐ and ☐

(1 mark) 4

**5** Finlay is investigating prime factors on his calculator.

He only uses the keys [2]; [3]; [5]; [×]; [=].

Each key can be used more than once.

Circle the products that Finlay **cannot** make using these prime factors.

20 35 28 21 36

(1 mark) 5

- Start to find prime factors by using the smallest possible prime number.

/5 Total for this page

# Multiplying and dividing by 10, 100 and 1,000

To achieve the higher score, you need to:

★ multiply and divide whole numbers and those involving decimals by 10; 100; and 1,000, giving answers up to three **decimal places**.

**1** a) 636 = 0.636 × ☐ b) 63.6 × ☐ = 6360

(1 mark)

**2** Each number in the second shape is $\frac{1}{10}$ of the number in the first shape.

Fill in the missing numbers on the first shape.

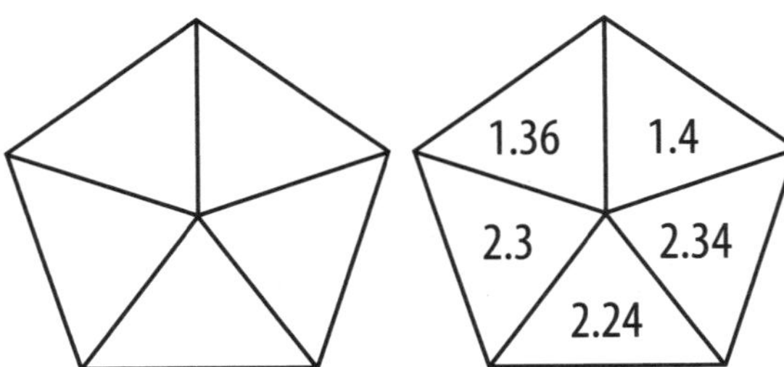


(1 mark)

**3** Alex is thinking of a number.

She multiplies it by 1,000 and then divides it by 10

The answer is 475

What was her number? ☐

(1 mark)

**4** ☐ × 200 = 65,000 ÷ 1,000

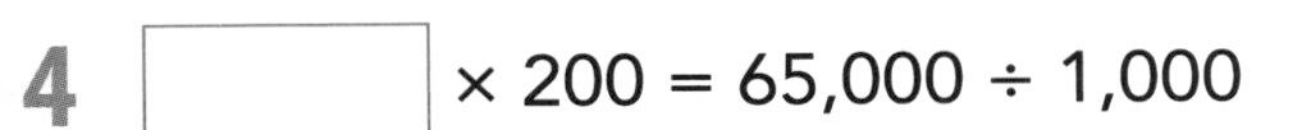

(1 mark)

**5** Anton marks out a shape using these measurements.

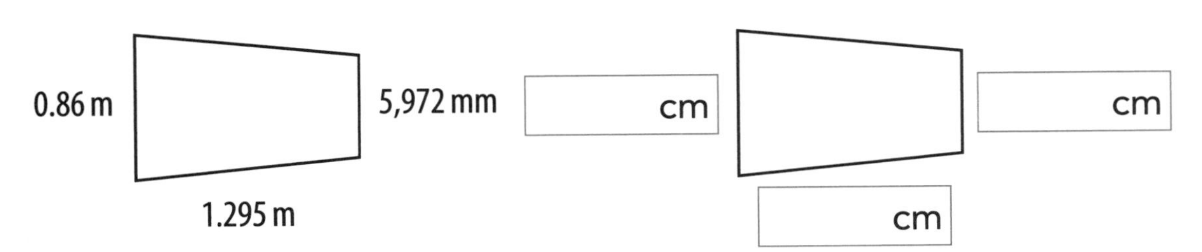


Write each of the measurements he uses in **centimetres**.

(1 mark)

**Top tip**

- Remember that multiplying and dividing by 10; 100; and 1,000 is key to converting between different units of measure, e.g. centimetres to metres, kilograms to grams.

/5 Total for this page

# Long multiplication

To achieve the higher score, you need to:

★ solve complex problems using the written method of **long multiplication**.

**1** There are 144 glitter pens in a box.

A shop orders 28 boxes.

The shop sells all the glitter pens at 75 pence each.

How much money does the shop take on glitter pens? ☐

(2 marks)

**2** Fill in the missing numbers in this calculation.

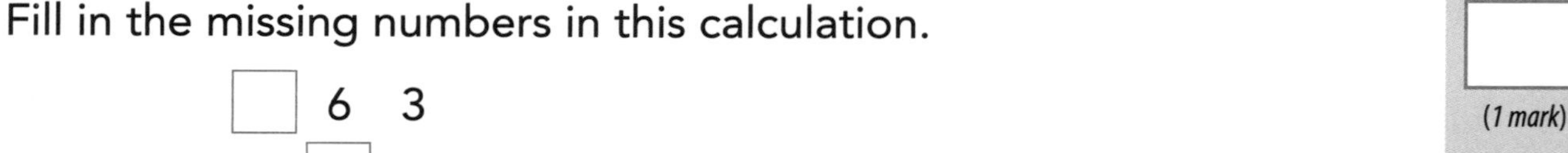

| | | | | | |
|---|---|---|---|---|---|
| | | | ☐ | 6 | 3 |
| × | | | | ☐ | 8 |
| | | 3 | 7 | 0 | 4 |
| | 3 | 2 | 4 | 1 | 0 |
| | 3 | 6 | 1 | 1 | 4 |

(1 mark)

**3** A map uses the scale 1 cm:175 km.

A route on the map is 36 cm.

What is the real distance of the route on the map? ☐ km

(1 mark)

**4** A bottle holds 635 ml of water.

Asha arranges bottles in rows of nine.

She makes six rows.

How many millilitres of water can all the bottles hold in **total**?

☐ ml

(2 marks)

**5** 27 15 33 16 24

Jack uses three of these cards for his calculation.

☐ × ☐ × ☐

The product is between 11,000 and 12,000

Which three cards does he use?

(1 mark)

**Top tip**

- Remember to include the placeholder 0 when multiplying by the tens digit.

/7 Total for this page

# Long division

To achieve the higher score, you need to:

- ★ use **long division** to solve problems and interpret remainders in context
- ★ use written division methods where the answer has up to two decimal places.

**1** 4,328 mugs are packed into boxes of 14

How many full boxes are there? [ ]

1 (1 mark)

**2** The area of a rectangle is 1,976 $cm^2$.

The length of the shorter side is 16 cm.

What is the length of the longer side? [ ] cm

2 (1 mark)

**3** A plane travels a distance of 7,644 miles in 13 hours.

On average, how many miles does it travel each hour? [ ] miles

3 (1 mark)

**4** Prize money of £6,490 is shared equally between 15 winners.

a) How much money will they get each? [ ]

4a (1 mark)

b) How much money will be left over? [ ]

4b (1 mark)

**5** Circle the numbers that represent the remainder after the division 2,408 ÷ 12

$\frac{1}{2}$ $\frac{2}{3}$ 8 9 $\frac{3}{4}$

5 (1 mark)

**6** A company wants to find the average cost of electricity per month over a two-year period.

In total, the company spent £9,564

What is the average cost of electricity per month? [ ]

6 (1 mark)

**Top tip**

- Always think carefully about the context of division problems when deciding on your final answer. If there is a remainder you may need to write this as a whole number, a fraction or a decimal. Remember, the answer to the calculation may not always be the final answer to the word problem.

/7 Total for this page

# Correspondence

To achieve the higher score, you need to:
★ solve **correspondence** problems in which *n* objects are connected to *m* objects.

**1** Alice is buying a present for Grandpa. She wants to buy a tie, scarf and hat.

She can choose from 4 different ties, 3 different scarves and 2 different hats.

How many different combinations can she make? [ ]

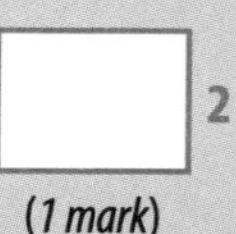

1 (1 mark)

**2** 1,944 people need 36 buses to take them to a football match.

972 people need 18 buses.

How many buses will 648 people need?

All buses carry the same number of people. [ ] buses

2 (1 mark)

**3** a)

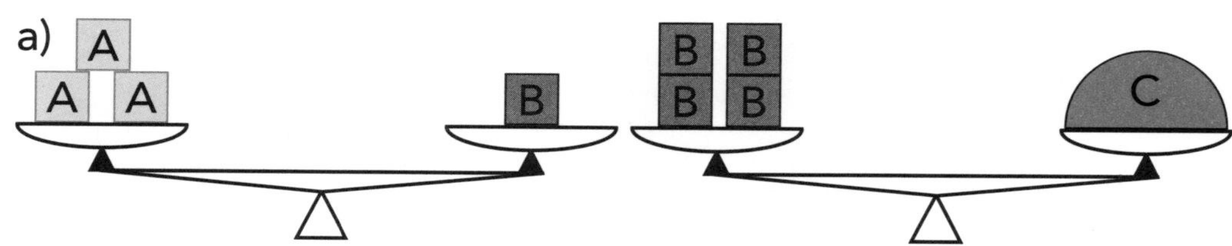


Complete the sentence to make it true.

Parcel C weighs the same as [ ] lots of Parcel A.

3a (2 marks)

b) Parcel D is also weighed.

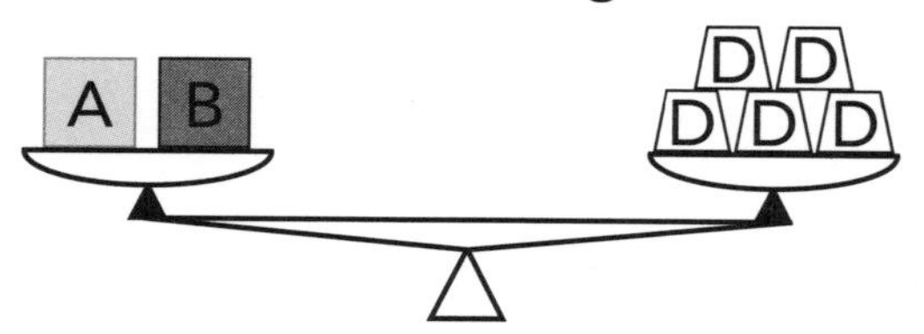


Complete the sentence to make it true.

Parcel C weighs the same as [ ] lots of Parcel D.

3b (2 marks)

**Top tip**

- In question 1, it is quicker to use a code rather than writing out the names of the items each time.

/6 Total for this page

# Order of operations

To achieve the higher score, you need to:
★ know the **order of operations** in which to carry out calculations involving the four operations and brackets.

**1** $180 - 73 + 9 \times 5 =$ ☐

(1 mark)

**2** ☐ $= 25 \times 3 + (30 - 10)$

(1 mark)

**3** Draw lines to match the calculation with the correct answer.

| Calculation | Answer |
| --- | --- |
| $200 - 50 \times 20$ | 1,000 |
| $125 + 75 \times 5$ | 3,000 |
| $2{,}500 + (4 \times 5^3)$ | 500 |
| $120 \times 50 \div 6$ | –800 |

(1 mark)

**4** 6 12 10 8 25

Use three different cards to make this calculation correct.

$16 + \square^2 = \square^2 - 4 \times \square$

(1 mark)

**5** Jayden says that the calculation $25 \times 8 \div 4$ can be done in any order.

Anna says that means that the calculation $25 + 8 \div 4$ can also be done in any order.

Are the children both correct?

Circle your answer. YES / NO

Explain your answer.

______________________________

______________________________

______________________________

(1 mark)

**Top tip**

- Read the whole calculation before you decide what you need to do first.

/5 Total for this page

# Ordering fractions

To achieve the higher score, you need to:
★ compare and order **fractions**, including fractions greater than 1.

**1** Write the symbol <, > or = to make these statements **true**.

$1\frac{5}{8}$ ☐ $1\frac{4}{7}$ $\frac{6}{9}$ ☐ $\frac{2}{3}$ $2\frac{9}{10}$ ☐ $2\frac{5}{6}$

(2 marks) 1

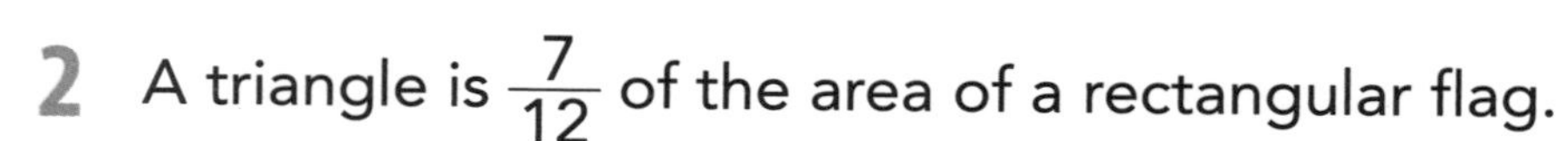

**2** A triangle is $\frac{7}{12}$ of the area of a rectangular flag.

A circle is $\frac{5}{9}$ of the area of another rectangular flag of the same size.

Which shape covers the greatest area? Tick (✓) your answer.

triangle ☐ circle ☐

(1 mark) 2

**3** Order these cards starting with the **smallest** fraction.

$\frac{7}{8}$ $\frac{5}{12}$ $1\frac{1}{2}$ $\frac{3}{4}$ $\frac{4}{5}$

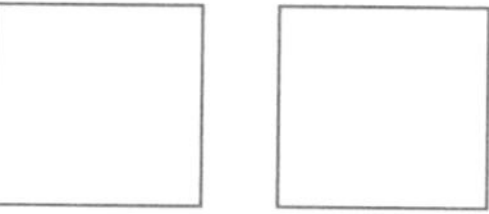

**smallest** **largest**

(1 mark) 3

**4** The shaded areas represent fractions of each shape.

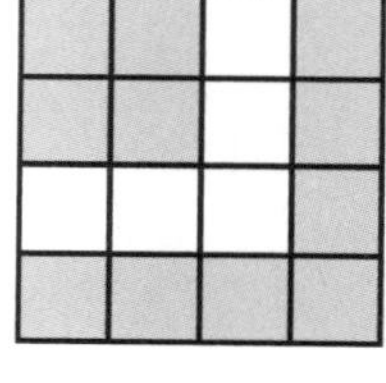

A

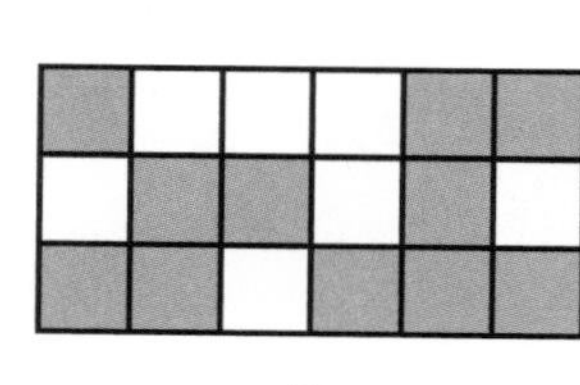

B

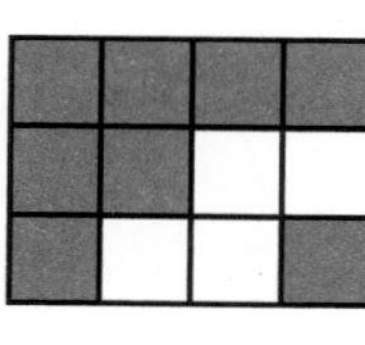

C

Write the letters in order starting with the **largest** fraction.

☐ ☐ ☐

**largest** **smallest**

(1 mark) 4

**Top tip**

- Use equivalent fractions to make denominators equal.

/5 Total for this page

# Adding and subtracting fractions

To achieve the higher score, you need to:
★ add and subtract fractions with different **denominators** and mixed numbers, using the concept of **equivalent fractions**.

**1** $\frac{3}{4} + \square + \frac{3}{8} = 2$

1 (1 mark)

**2** Harry eats $\frac{2}{3}$ of his pizza. Charlotte eats $\frac{7}{9}$ of her pizza.

a) What fraction of pizza has been eaten in **total**? ☐

2a (1 mark)

b) What **total** fraction of pizza is left? ☐

2b (1 mark)

**3** A large carton holds $2\frac{5}{8}$ litres of juice.

$\frac{2}{3}$ litre of juice is poured from the carton.

How much juice is **left** in the carton? ☐

3 (1 mark)

**4** 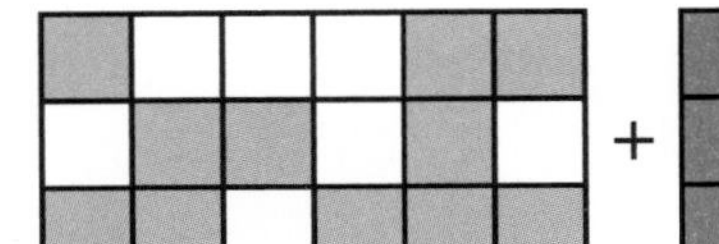 + 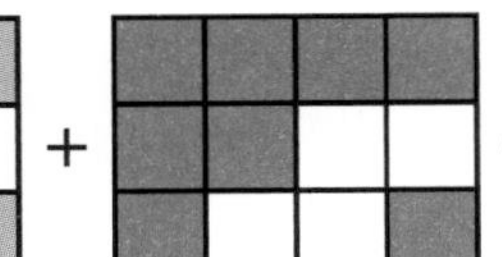 = ☐

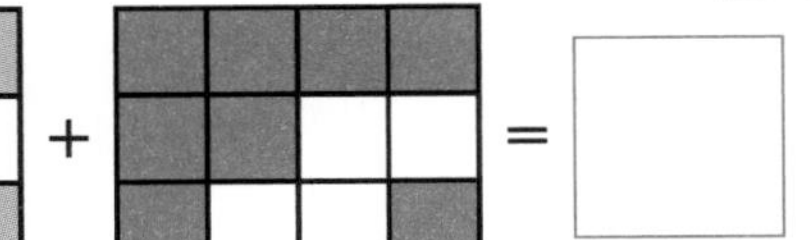

4 (1 mark)

**5** A café sells **five** pieces from each of the cakes below.

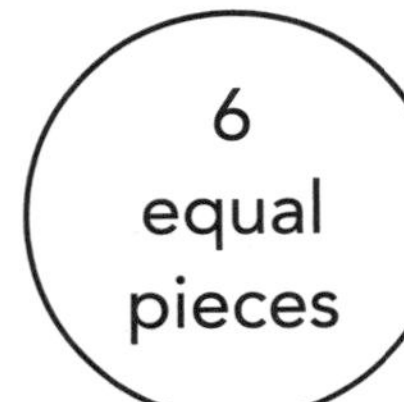

6 equal pieces | 8 equal pieces | 12 equal pieces

How much cake has been sold in total?

Give your answer in its **simplest form**. ☐ cakes

5 (2 marks)

**Top tip**
- Use equivalent fractions to make denominators equal.

/7 Total for this page

# Multiplying fractions

To achieve the higher score, you need to:
- ★ multiply **proper fractions** and mixed numbers by whole numbers
- ★ multiply simple pairs of proper fractions, writing the answer in its simplest form.

**1** $\frac{3}{5} \times 6 =$ ☐

(1 mark)

**2** $\frac{4}{5} \times \frac{7}{10} =$ ☐

(1 mark)

**3** A recipe uses $\frac{5}{8}$ kg of flour to make 10 cakes.

How much flour is needed to make 30 cakes? ☐ kg

(1 mark)

**4** What is the area of this rectangle? ☐ $m^2$

$\frac{2}{3}$ m

$\frac{3}{4}$ m

(1 mark)

**5** What is $\frac{3}{5}$ of $2\frac{1}{2}$ litres ? ☐ litres

(1 mark)

**6** Kris has a drawer full of pens, pencils and erasers in his office. The pie chart compares the different numbers of each item in the drawer.

$\frac{3}{10}$ of the pens are blue.

What **fraction** of the **total** number of items in the drawer are blue pens? ☐

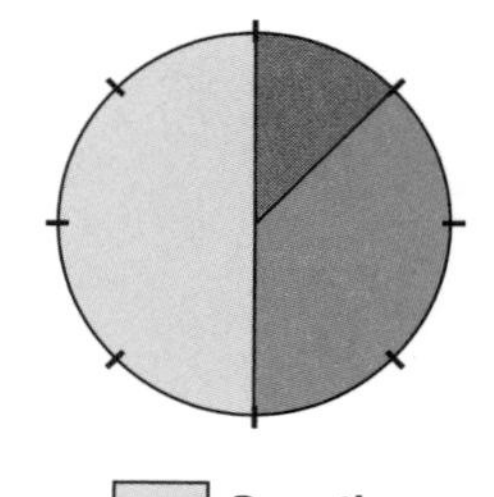

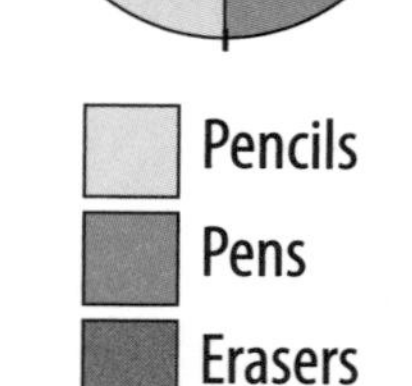


(1 mark)

**Top tip**

- Remember to write answers in their simplest form using your knowledge of equivalent fractions.

/6

Total for this page

# Dividing fractions

To achieve the higher score, you need to:
★ divide **proper fractions** by a whole number.

**1** $\frac{5}{8} \div 3 =$ ☐

*(1 mark)* 1

**2** $\frac{3}{20} = \frac{3}{4} \div$ ☐

*(1 mark)* 2

**3** Gran cuts her chocolate cake into equal pieces.

She then cuts one of these pieces into 5 smaller equal pieces.

What **fraction** of the whole cake is one of these smaller pieces? ☐

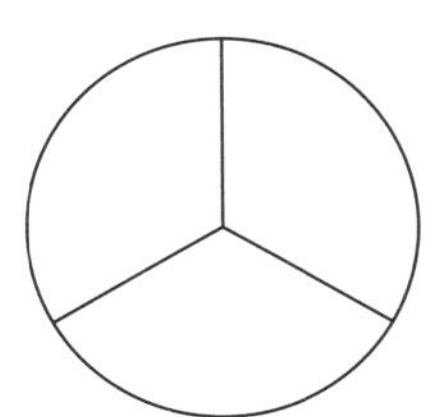

*(1 mark)* 3

**4** This pie chart compares the different items of stationery that Abdul buys for the office.

Abdul shares the pens equally between 6 of his staff.

What **fraction** of the total number of items does each person's share of pens represent? ☐

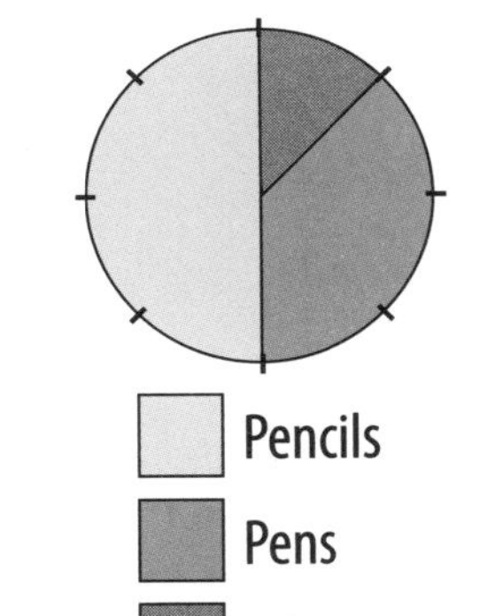


*(2 marks)* 4

**5** Zac cuts a $6\frac{3}{4}$ m ribbon into 9 equal pieces.

What **fraction** of a metre is each piece? ☐ m

*(1 mark)* 5

**Top tip**

- Remember that dividing by a number such as 2 is the same as finding $\frac{1}{2}$.

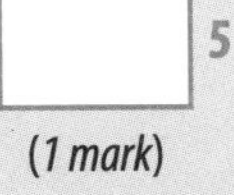

/6

*Total for this page*

# Changing fractions to decimals

To achieve the higher score, you need to:
★ use division to calculate decimal fraction equivalents.

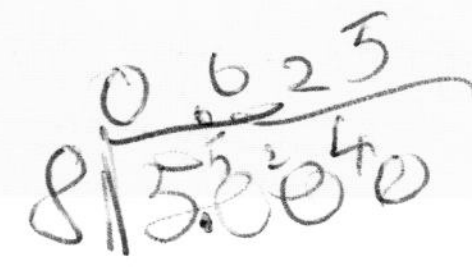

**1** Write $4\frac{3}{4}$ as a decimal. 4.75 (1 mark)

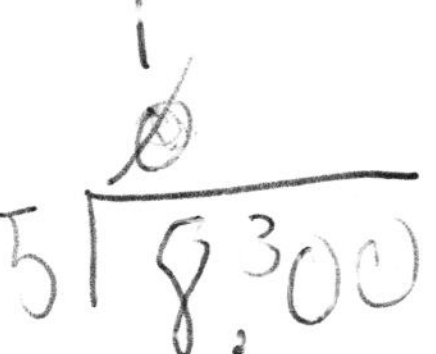

**2** Write $\frac{5}{8}$ as a decimal. 0.625 (1 mark)

**3** Write $3\frac{2}{3}$ as a decimal rounded to **two** decimal places. 3.67 (1 mark)

**4** Circle the values that are equivalent to the answer to this division.

35 ÷ 4 = 8 r3

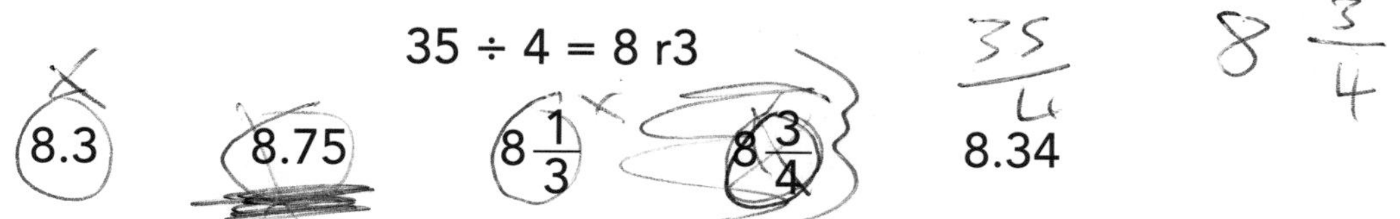

8.3 8.75 $8\frac{1}{3}$ $8\frac{3}{4}$ 8.34

(1 mark)

**5** Jamie completes this calculation: 47 ÷ 5 = 9 r2

Write the remainder as a fraction and a decimal.

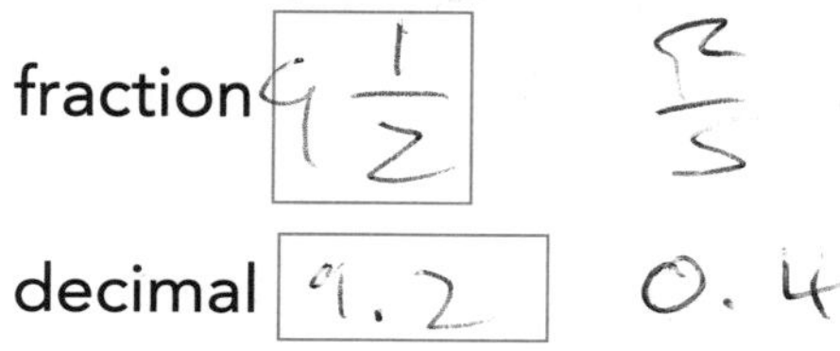

fraction

decimal

(1 mark)

**6** Write the **remainder** to this calculation as a fraction and a decimal.

35 ÷ 8 =

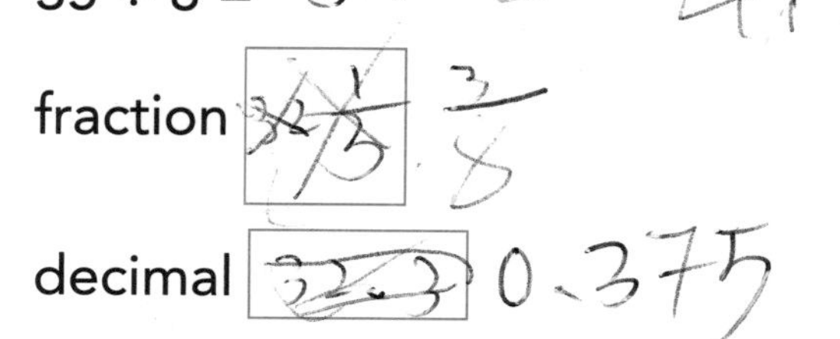

fraction

decimal

(1 mark)

**Top tip**

- To turn a fraction into a decimal, divide the numerator by the denominator.

/6 Total for this page

# Rounding decimals

To achieve the higher score, you need to:

★ round decimals with up to three **decimal places** to the nearest whole number, or nearest one, or two, decimal places.

**1** Round 36.465 to:

a) **one** decimal place ☐ b) **two** decimal places ☐

1 ☐ *(1 mark)*

**2** Write your answer rounded to **two** decimal places.

575 ÷ 9 = ☐

2 ☐ *(1 mark)*

**3** This bar chart shows the number of passengers passing through an airport's terminals in one month.

All numbers are rounded to one decimal place.

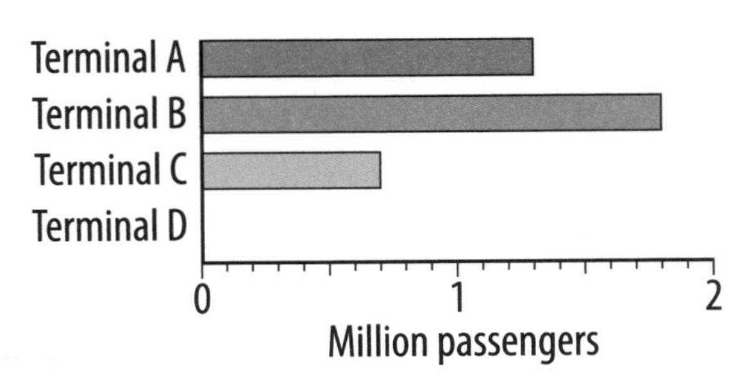


| Terminal A | Terminal B | Terminal C | Terminal D |
|---|---|---|---|
| 1.273 million | 1.846 million | 0.652 million | 1.459 million |

Draw the bar to represent the passengers passing through Terminal D in that month.

3 ☐ *(1 mark)*

**4** A mass is rounded to the nearest tenth of a kilogram.

The rounded mass is 2.8 kg.

Circle all the possible original masses.

2.77 kg 2.749 kg 2.85 kg 2.849 kg 2.809 kg

4 ☐ *(1 mark)*

**5** A capacity is rounded to two decimal places.

The rounded capacity is 3.04 litres.

Circle all the values that **cannot** be the original capacity.

3.0483 litres 3.039 litres 3.0401 litres

3.0449 litres 3.0349 litres

5 ☐ *(1 mark)*

**Top tip**

- Remember that the rules for rounding whole numbers also apply to rounding decimals.

☐ /5 *Total for this page*

# Adding and subtracting decimals

To achieve the higher score, you need to:

★ add and subtract decimals with up to three **decimal places**, and numbers with different numbers of decimal places.

**1**

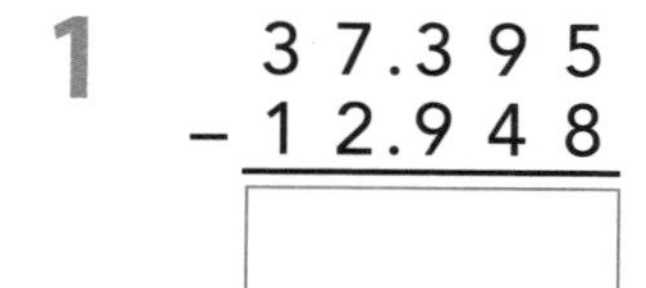

(1 mark)

**2** Choose two numbers from the shape each time to make these statements **true**.

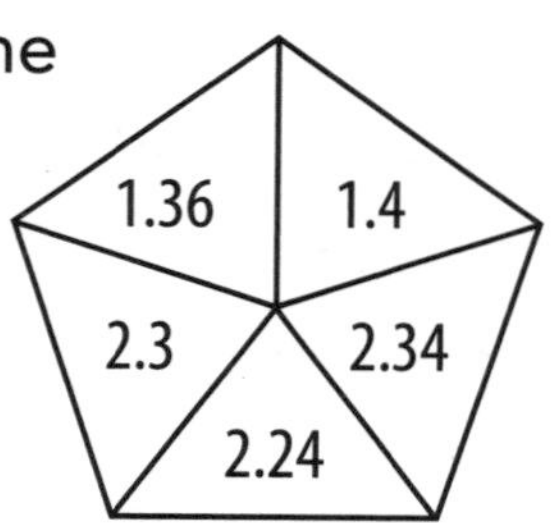


2a (1 mark)

2b (1 mark)

**3** Kim orders some materials for her garden.

1.3 kg gravel
2.75 kg sand
0.368 kg soil

a) How much **greater** is the mass of the sand than the soil? ____ kg

3a (1 mark)

b) What is the **total** mass of gravel, sand and soil? ____ kg

3b (1 mark)

**4** 12.46 + 13.538 + ____ = 40.498

4 (1 mark)

★ **5** Bilal pours 1.675 litres of water into a 2.5 litre jug.

He adds 0.24 litres of orange to make squash.

How much **more** squash can the jug hold? ____ litres

5 (2 marks)

**Top tip**

- Line up the decimal points when you write out a calculation.

/8 Total for this page

# Multiplying decimals

To achieve the higher score, you need to:

★ multiply one-digit numbers with up to two **decimal places** by one-digit and two-digit whole numbers.

**1** [ ] × 8 = 6.4

0.8 × [ ] = 64

[ ] × 8 = 0.64

(1 mark) 1

**2** 3 × [ ] = 0.4 × 6

(1 mark) 2

**3** Sara multiplies 0.09 by a single-digit number.

a) Circle the numbers that **cannot** be the answer to her calculation.

0.36    3.6    0.45    8.1    0.56

(1 mark) 3a

b) Explain your thinking.

______________________________

______________________________

______________________________

(1 mark) 3b

**4** Fill in the missing boxes to make these multiplications correct.

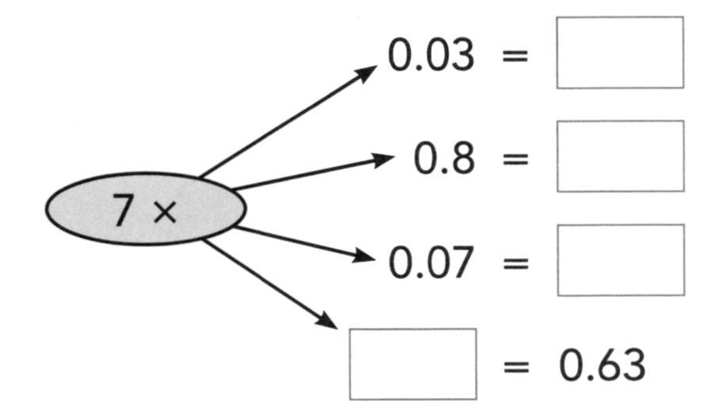


(2 marks) 4

**5** A rectangle has the dimensions 0.3 m by 0.09 m.

Calculate the area of the rectangle. [ ] 

(1 mark) 5

**Top tip**

- Think about how the place value of decimal multiplication is related to known multiplication facts.

/7 Total for this page

# Percentages

To achieve the higher score, you need to:

- ★ solve problems involving the calculation of **percentages** and use percentages for comparison
- ★ compare percentages and fractions including tenths, fifths, halves and quarters.

**1** Freddy wins 25% of £1,500 prize money.

Jahla wins 40% of £900 prize money.

Who wins the most money? ____________

How much more do they win? ________

*(1 mark)* 1

**2** Jan buys a TV in the sale. The sale price is £448

What was the original price of the TV? £ ______

Sale: 30% off

*(2 marks)* 2

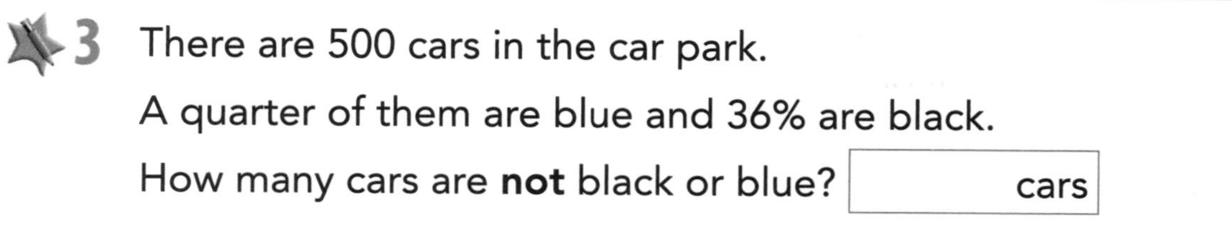

**3** There are 500 cars in the car park.

A quarter of them are blue and 36% are black.

How many cars are **not** black or blue? ______ cars

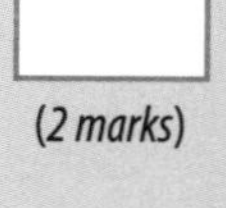

*(2 marks)* 3

**4** Tick (✓) the shapes that are shaded $\frac{2}{5}$ light green and 60% dark green.

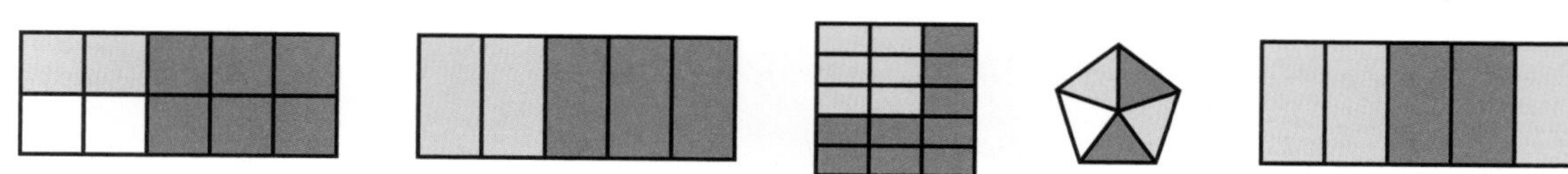

*(1 mark)* 4

**5** Amy says that the fraction $\frac{24}{40}$ is equal to 60%.

Philip disagrees because he always writes percentages as fractions with a denominator of 100

Stefan disagrees because the denominator 40 is not a factor of 100

Is Amy correct? Circle your answer. YES / NO

Explain how you know.

______________________________________________

______________________________________________

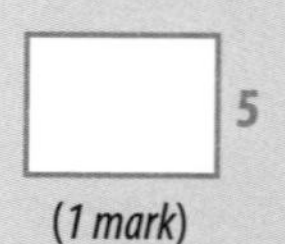

*(1 mark)* 5

**Top tip**

- Think about what percentages you know and how to find the percentage you are being asked for.

/7

*Total for this page*

# Ratio

To achieve the higher score, you need to:
★ solve **ratio** problems where missing values can be found by using multiplication and division facts.

**1** In a fruit salad, 12 strawberries are used for every 5 peaches.

Flora uses 85 of these fruits altogether to make a large fruit salad.

How many strawberries does she use? ☐

1 *(1 mark)*

**2** Jo draws a square with a base of 4.5 cm.

She then scales the base by a whole number to draw a larger square.

Circle the base lengths that she **cannot** make.

27 cm 45 cm 54 cm 84 cm 90 cm

2 *(1 mark)*

**3** The ratio of fish to frogs in a small pond is 5:2

There are 14 frogs.

How many fish and frogs are there in total? ☐

3 *(1 mark)*

**4** The picture shows one part of a repeating pattern of rectangular and circular tiles.

4 *(2 marks)*

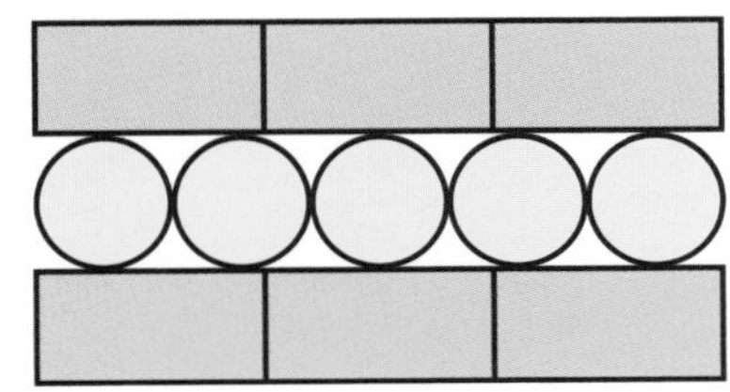

Complete the table using the same ratio as shown in the picture.

| Rectangular tiles ▭ | 18 | | 66 | 120 | |
|---|---|---|---|---|---|
| Circular tiles ○ | | 45 | | | 150 |

**Top tip**

- Ratio describes how something is divided and tells you the relative sizes of the shares.

/5 Total for this page

# Proportion

To achieve the higher score, you need to:
★ solve problems involving the relative sizes of two quantities.

**1** There are 32 children in the class. They are all 10 or 11 years old.
12 children are 11 years old.
What **proportion** of the class are 10 years old?

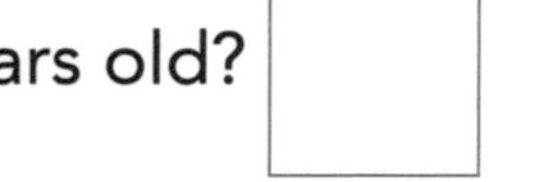

(1 mark)

**2** A pet shop sells 150 g of dog biscuits for £1.20
How many grams of dog biscuits can be bought for £4.20? g

(1 mark)

**3** The same pet shop sells 240 g of bird seed for £1.60

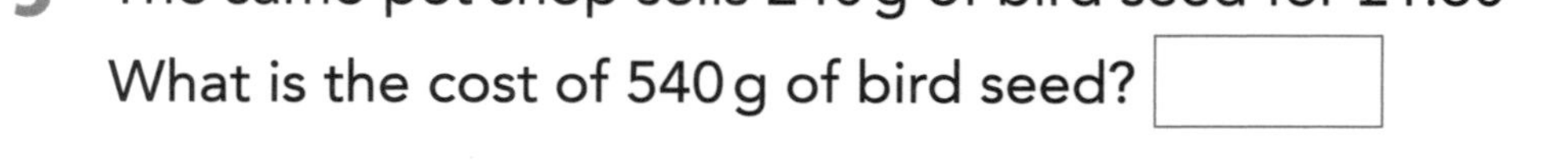

What is the cost of 540 g of bird seed?

(1 mark)

**4** A distance on a map is measured as 22.5 cm.
The real distance is 675 km.
Fill in the scale that is used on the map.
1 cm = km

(1 mark)

**5** The scale on a map is 1 cm = 150 m.
Nasser measures the distance on the map between his house and the swimming pool as 4.5 cm.
How far is the swimming pool from his house in real life?
Distance = m

(1 mark)

★ **6** 6 small bricks have the same mass as 5 large bricks.

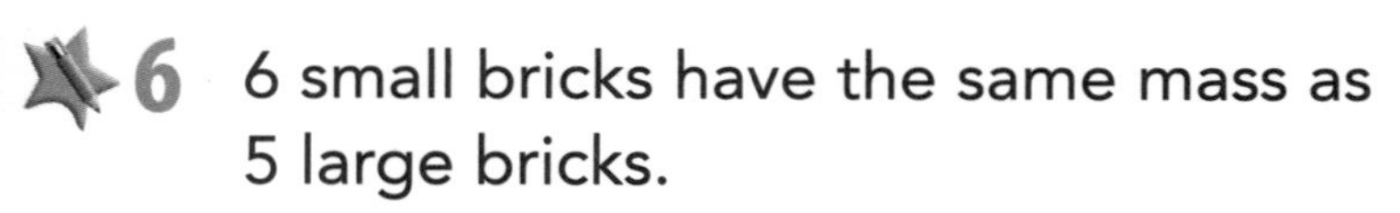

The mass of one small brick is 3.5 kg.
What is the mass of one large brick?

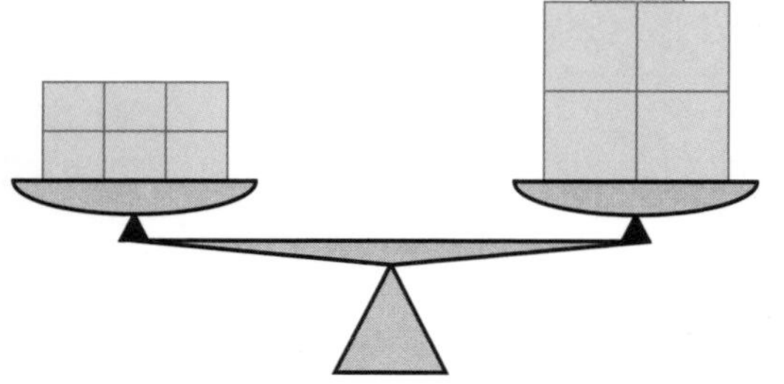

kg

(2 marks)

**Top tip**
• Remember that proportion can be used to compare part to whole. Always try to identify the whole first.

/7
Total for this page

# Scaling problems

To achieve the higher score, you need to:
- ★ solve scaling problems, including scaling by simple fractions
- ★ solve problems where the **scale factor** is known or can be found.

**1** Gita draws a square with a base of 4.5 cm.

She then draws a smaller square with a base that is $\frac{2}{3}$ of this length.

Draw her new square on the grid.

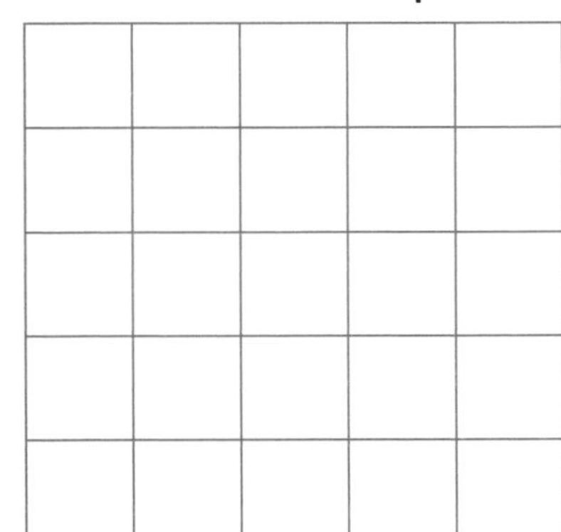

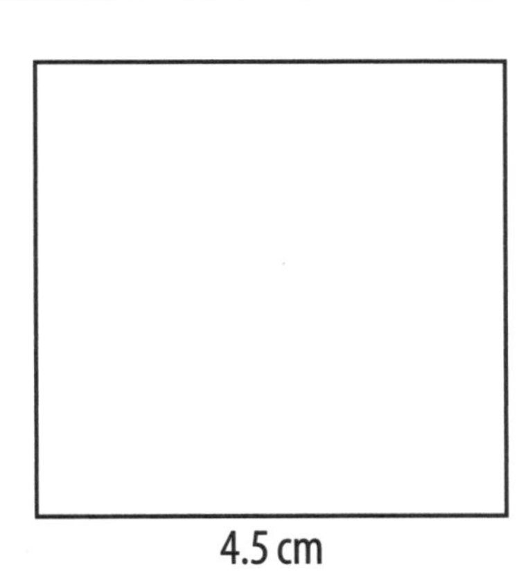

*(1 mark)*

**2** Kayla enlarges this triangle using a scale factor of 4

What are the dimensions of the larger triangle?

cm and cm

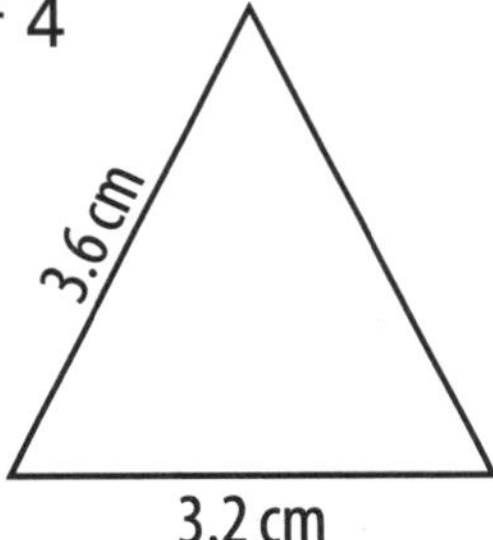


2 *(1 mark)*

**3** 36 identical boxes have a total mass of 180 kg.

What is the mass of 24 of these boxes?

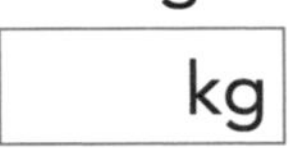

*(1 mark)*

**4** 7.8 cm A 14 cm     x B 21 cm     Not to scale

Rectangle B is an enlargement of rectangle A.

Calculate the length of side x. cm

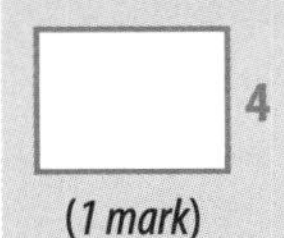


*(1 mark)*

- Draw accurately using a ruler to gain full marks.

/4

*Total for this page*

# Unequal sharing

To achieve the higher score, you need to:

★ solve problems involving unequal sharing and grouping using your knowledge of fractions and multiples.

**1** A shop sells books, magazines and newspapers in the ratio 3:5:10

It sold 90 books on Saturday.

a) How many magazines did it sell? [ ] *(1 mark)* 1a

b) How many **more** newspapers than books did it sell? [ ] *(1 mark)* 1b

**2** There are 120 marbles in a jar.

Mo takes three times as many as Darren.

Belle takes double the number that Darren has.

How many marbles does each child take?

Mo [ ] Darren [ ] Belle [ ] *(1 mark)* 2

**3** Jon collects £2 coins and £1 coins in the ratio 5:9

He has 280 coins in total.

a) How many of each coin does Jon have?

[ ] £2 coins and [ ] £1 coins *(1 mark)* 3a

b) What is the **total** value of his coins? £ [ ] *(1 mark)* 3b

c) Binoy also collects £2 coins and £1 coins, but in the ratio 13:7

He also has 280 coins in total.

How much **more** money does he have than Jon? £ [ ] *(2 marks)* 3c

- Remember to check the ratios first. If no ratio is given, you must identify it.

/7 Total for this page

# Algebra

To achieve the higher score, you need to:

★ express missing number problems using **algebra**.

**1** Henry thinks of a number and divides it by 4

He then adds 5

Circle the algebraic rule that Henry has used.

$4n + 5$ $(4 \div n) + 5$ $(n \div 4) + 5$ $\frac{n}{5} + 4$

1 (1 mark)

**2** Use the algebraic rule $4n - 3$ to fill in the missing boxes.

| Value of $n$ | Answer |
|---|---|
| 16 | ☐ |
| ☐ | 81 |
| 4.2 | ☐ |

2 (1 mark)

**3** The large shapes are made up using these smaller squares and triangles. $a$ $b$

Describe each shape algebraically using $a$ and $b$.

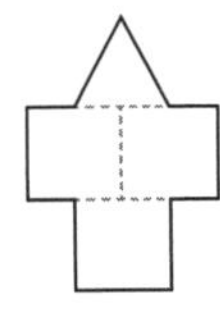

a) ______________ b) ______________

3a (1 mark)

3b (1 mark)

**4** Kiki uses an algebraic rule to plot these coordinates.

Complete her rule.

$y =$ ______________

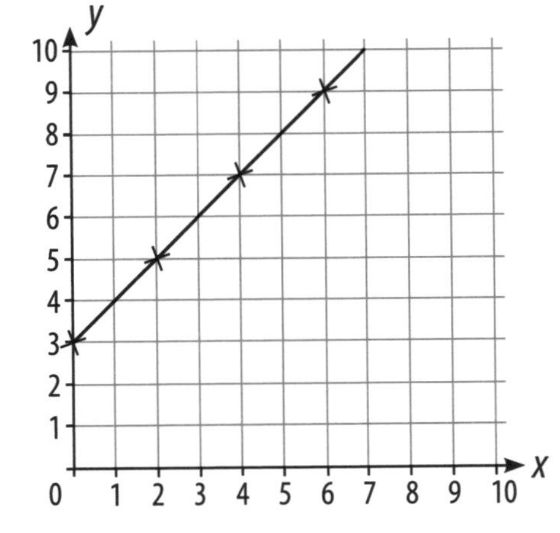


4 (1 mark)

**Top tip**

- Remember to check your answers using the inverse operation.

/5 Total for this page

# Using formulae

To achieve the higher score, you need to:
★ use and write simple **formulae** using algebra.

**1** The radius of a circle can be calculated using the formula $r = d \div 2$

a) What is the radius of this circle?

[ cm ]

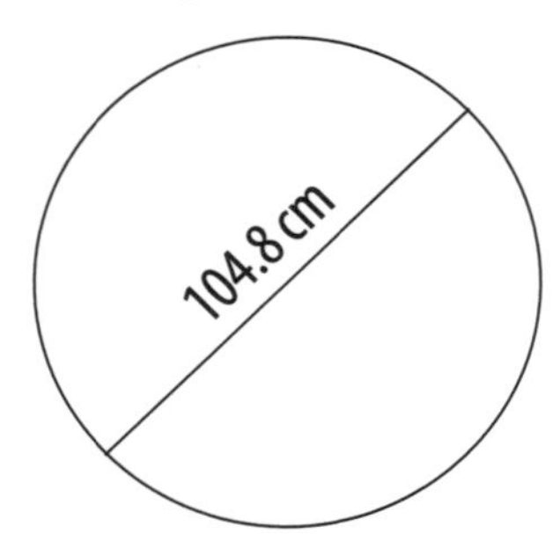


(1 mark) 1a

b) What is the radius of this circle?

[ cm ]

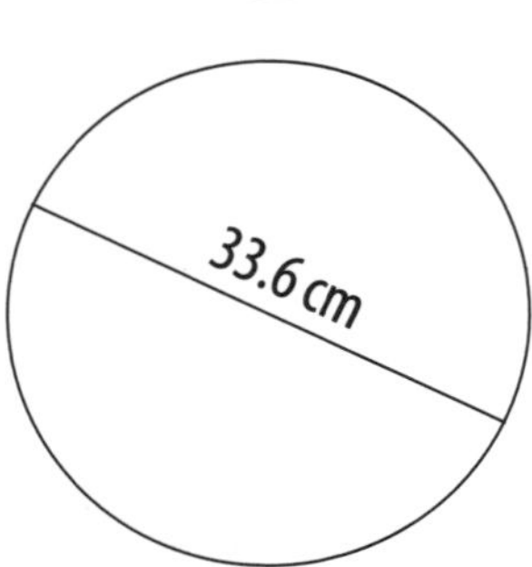

(1 mark) 1b

**2** Vicki uses the formula $b = 360° - a - c$.

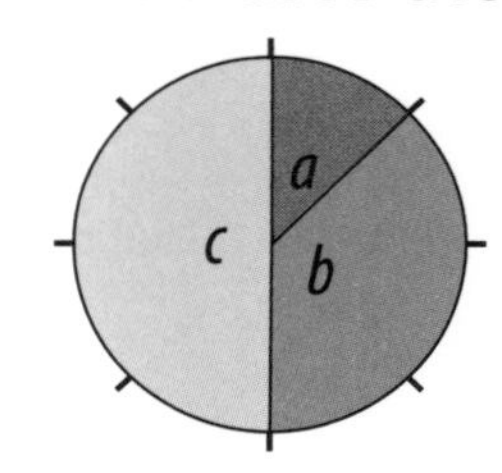


Rewrite the formula in degrees using the information given here.

$b = 360° -$ [ ]° $-$ [ ]°

(1 mark) 2

**3** The following formula is used to convert temperatures from degrees Fahrenheit (°F) to degrees Celsius (°C):

$°C = (°F - 32) \times \frac{5}{9}$

Use the formula to complete the table.

| Fahrenheit (°F) | 77 | 50 | 59 |
|---|---|---|---|
| Celsius (°C) | | | |

(2 marks) 3

**Top tip**

- Remember to check your answers using the inverse.

/5 Total for this page

# Solving equations

To achieve the higher score, you need to:
★ find a pair of numbers that satisfies an **equation** with two unknowns
★ calculate possible combinations of two **variables**.

**1** $m = 120 \div n$

$n^3 = 70 - 6$

What are the values of ***m*** and ***n***?

$m =$ ☐ $n =$ ☐

(1 mark) 1

**2** Find the values of the shapes in both these equations.

⬠ × ☐ = 60

⬠ + ⬠ – 19 = 11

☐ = ☐

⬠ = ☐

(1 mark) 2

**3**

| A $54\,m^2$ | B $30\,m^2$ |
|---|---|

The area of a third rectangle (C) can be expressed as:

Area C = Area A – Area B

What are the possible dimensions of rectangle C when the measurements are whole metres?

☐ m × ☐ m

(1 mark) 3

**4** $2a + 3b - c = 18$

The value of ***c*** is double the value of ***a***.

All values are whole numbers.

Find the missing value ***b***.

$b =$ ☐

(1 mark) 4

**Top tip**

- Be systematic. Calculate what you can first. Rewrite the equation putting your calculated values in place to check if your values are correct.

/4

Total for this page

# Linear number sequences

To achieve the higher score, you need to:
★ generate, describe and complete **linear number sequences**.

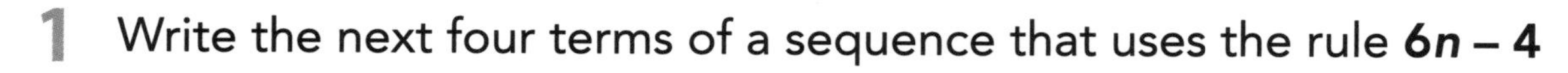

**1** Write the next four terms of a sequence that uses the rule $6n - 4$

| 2 | | | | |
|---|---|---|---|---|

(1 mark) 1

**2** Circle the rule that describes this sequence: 3 7 11 15 19

$3n + 4$  $4n + 3$  $4n + 1$  $4n - 1$

(1 mark) 2

**3**

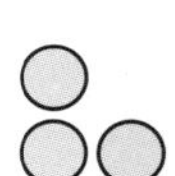
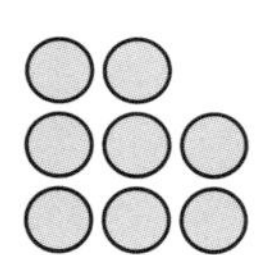
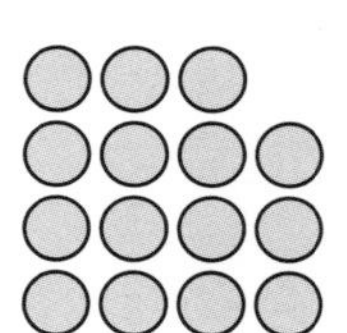
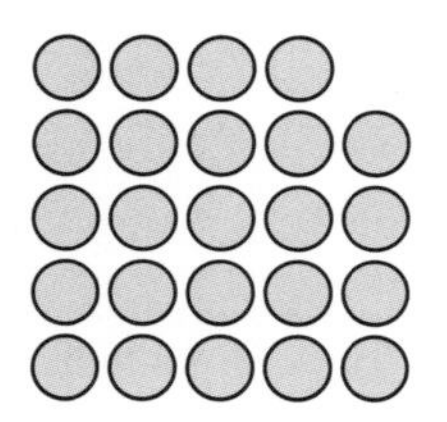

?

a) Write the number of circles that are needed to draw the missing pattern in this sequence. ☐

(1 mark) 3a

b) How many circles are needed to draw the 10th pattern of circles in the sequence? ☐

(1 mark) 3b

**4**

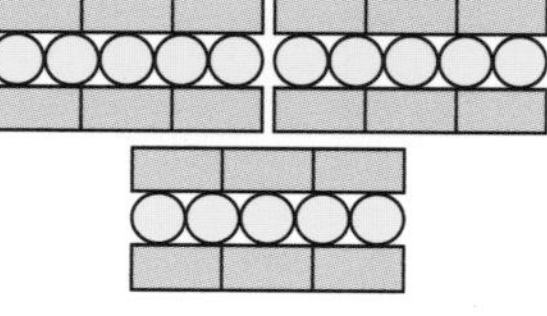
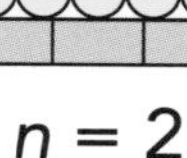
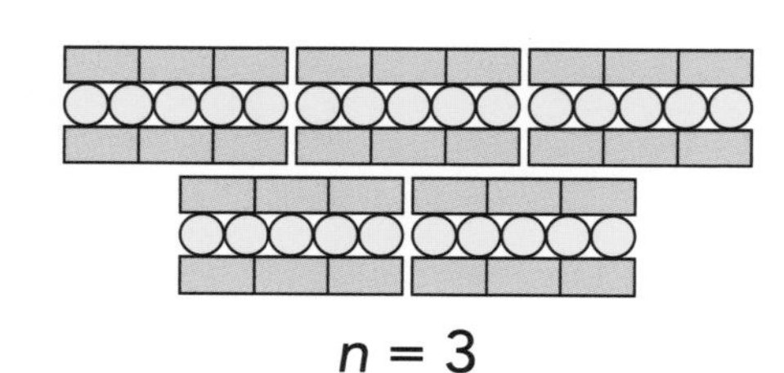

$n = 1$  $n = 2$  $n = 3$

The number of circles in this sequence can be expressed using the rule $c = 10n - 5$

Complete the rule below using $n$ to express the number of rectangles each time.

$r = \square n - 6$

(1 mark) 4

**Top tip**

- Remember that $n$ is used to represent the position in the sequence.

/5
Total for this page

# Measures

To achieve the higher score, you need to:
★ solve problems involving the calculation and conversion of units of measure, using up to three **decimal places.**

**1** A 25 kg sack of flour is divided into smaller bags of 525 g.

a) How many small bags of flour can be made? [ ]

b) What is the mass of flour left in the sack? [ kg]

1a (1 mark)
1b (1 mark)

**2** Kate has an equal number of coins in three bags.

She exchanges them at the bank for **two** notes.

Circle the number of coins that could be in each bag.

10 15 20 25 40

2 (1 mark)

**3** Toby uses these bricks to build a wall. A space of 0.8 cm is needed between bricks for cement.

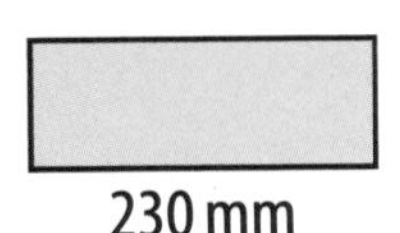


The wall is 21 bricks long.

Calculate the actual length of the wall. [ m]

3 (2 marks)

**4** A swimming pool is filled with water at a rate of 26.485 litres every 210 seconds.

Calculate the amount of water that goes into the pool in 14 minutes.

[ litres]

4 (2 marks)

## Top tip

- Remember to check the units of measurement each time and convert any that are not the same.

/7 Total for this page

# Converting metric units

To achieve the higher score, you need to:
- ★ convert between different units of **metric** measure
- ★ convert measurements of **length, mass, volume** and **time** from a smaller unit of measure to a larger unit, and vice versa, using up to three **decimal places.**

**1** Complete this conversion table.

| km | | 0.45 | | |
|---|---|---|---|---|
| m | | | 143 | |
| cm | 1,210 | | | 13,700 |

1 (2 marks)

**2** Omar buys 2.34 m of rope.

Petra buys 98.3 cm more rope than Omar.

What is the length of Petra's rope? ______ m

2 (1 mark)

**3**

| <500g | <1,000g | <1,500g | <2,000g | <2,000g |
|---|---|---|---|---|
| 3 days | Next day | Next day | Next day | 3 days |
| £1. 48 | £3.20 | £4.95 | £5.45 | £3.90 |

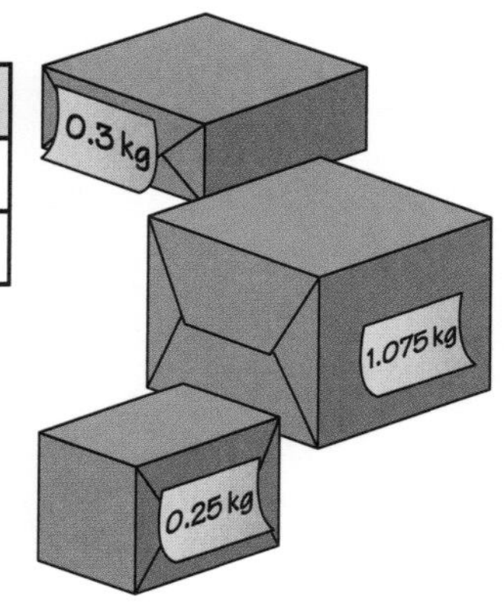


Gran puts all three presents into a larger box.

What is the cost of sending the parcel with *Next day* delivery? ______

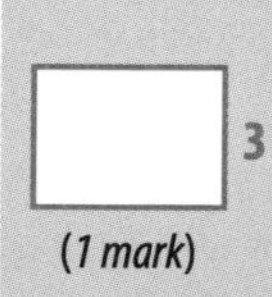

3 (1 mark)

**4** A digital TV box can store 120 hours of programmes.

The average length of different programmes is shown here.

| Films | Sports | Nature |
|---|---|---|
| 1.5 hours | $2\frac{1}{4}$ hours | 75 minutes |

Amy has recorded 12 films, 9 sports and 15 nature programmes.

She calculates that she has approximately half of the recording space left on her digital TV box.

Do you agree with Amy? Circle your answer. YES / NO

Explain your answer. ______________________________

______________________________

4 (1 mark)

/5 Total for this page

# Metric units and imperial measures

To achieve the higher score, you need to:

- ★ understand and use approximate equivalences between metric units and common **imperial units** such as inches, pounds and pints
- ★ convert between miles and kilometres.

**1** Complete this conversion table.

30 cm ≈ 1 foot

| centimetres | | 660 | |
|---|---|---|---|
| feet | 7.5 | | $5\frac{1}{4}$ |

1 (1 mark)

**2** 5 miles ≈ 8 km

The distance between two cities is 450 miles.

What is this distance in kilometres? ______ km

2 (1 mark)

**3** 1 kilogram is approximately 2.2 pounds.

What is the mass on the scales in kilograms?

______ kg

3 (1 mark)

**4** A pint is approximately 568 ml.

Jake pours 3.5 pints of milk into a 2 litre jug.

How much **more** milk can be poured into the jug? ______ ml

4 (1 mark)

**5** 1 inch is approximately 2.5 cm. There are 12 inches in a foot.

Callum is 1.92 metres tall.

Megan is 5 foot 7 inches tall.

How much taller is Callum? ______ cm

5 (1 mark)

**Top tip**

- Estimate your answers before you do the calculations.
- ≈ means approximate.

/5

*Total for this page*

# Perimeter and area

To achieve the higher score, you need to:

★ measure and calculate the **perimeter** and **area** of composite rectilinear shapes in centimetres and metres.

**1** Draw a rectangle with the **same** perimeter as this shaded shape.

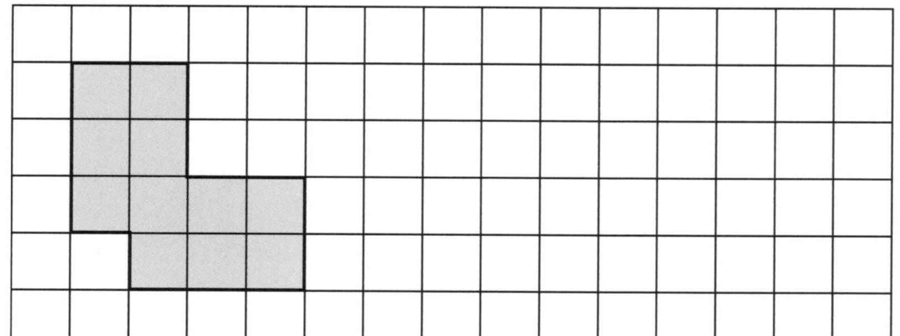

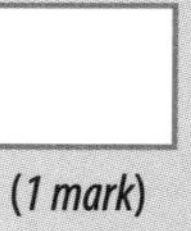

1

(1 mark)

**2** Shade in **two more** squares so that the perimeter of this shape stays the **same**.

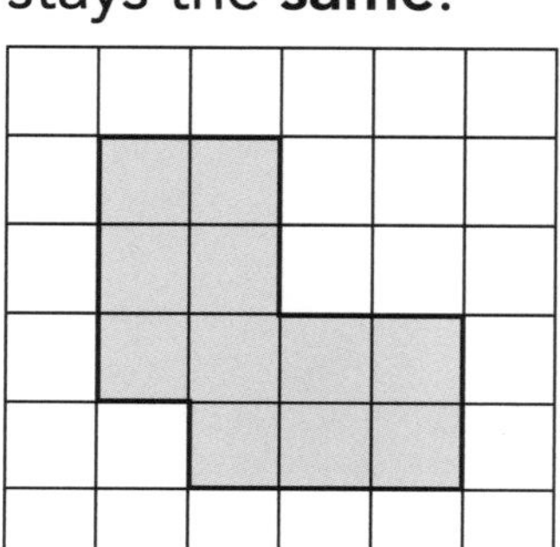

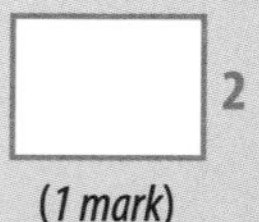

2

(1 mark)

**3** Paul draws 16 of these squares to make a grid for a board game.

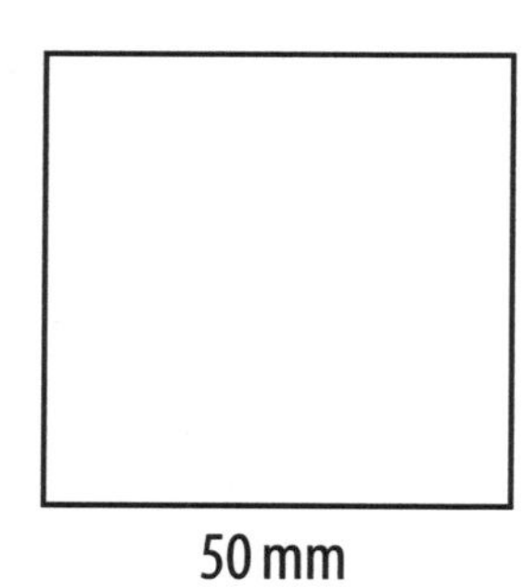


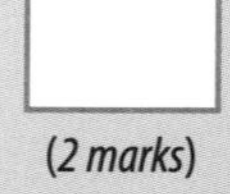

3

(2 marks)

His grid can be rectangular or square in shape.

Calculate the **shortest** and the **longest** possible perimeter his grid can have.

shortest = [ ] cm

longest = [ ] cm

/4

Total for this page

**4** Calculate the area of this shape. ☐ cm²

(1 mark)

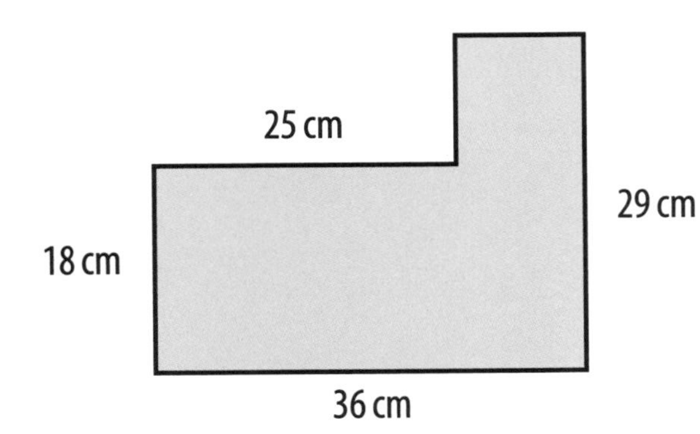


**5** The perimeter of this shape is 48 cm.

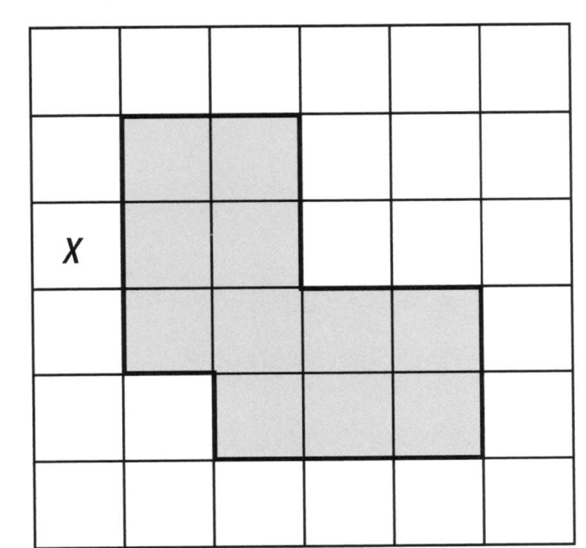


What is the length of the side marked $x$? ☐ cm

(2 marks)

**6** A square tile measures 10 cm by 10 cm.

A rectangular tile is 2 cm **longer** and 4 cm **narrower** than the square tile.

What is the **difference in area** between the two tiles? ☐

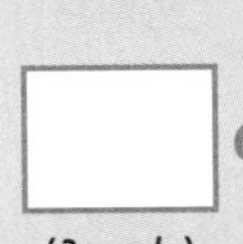

(3 marks)

**Top tip**

- Write down any lengths you find on the shape so that you remember to include them in your calculation.

/6

Total for this page

# Area of parallelograms and triangles

To achieve the higher score, you need to:

★ calculate the area of parallelograms and triangles.

**1** a) Tick (✓) the triangle that has the **greatest** area.

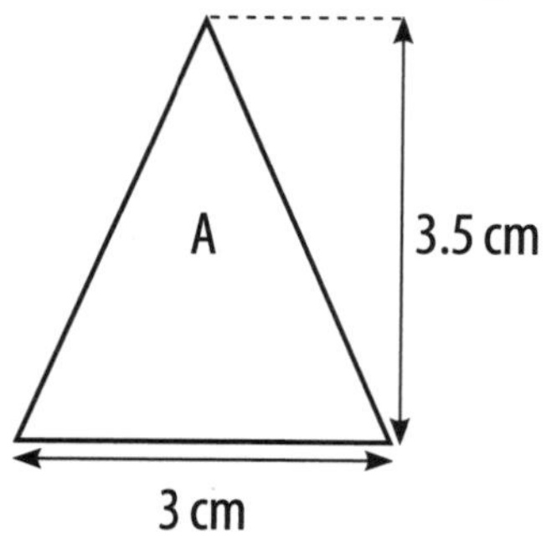


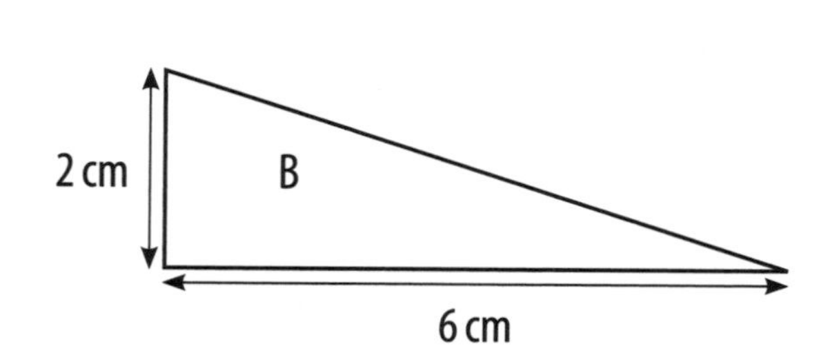


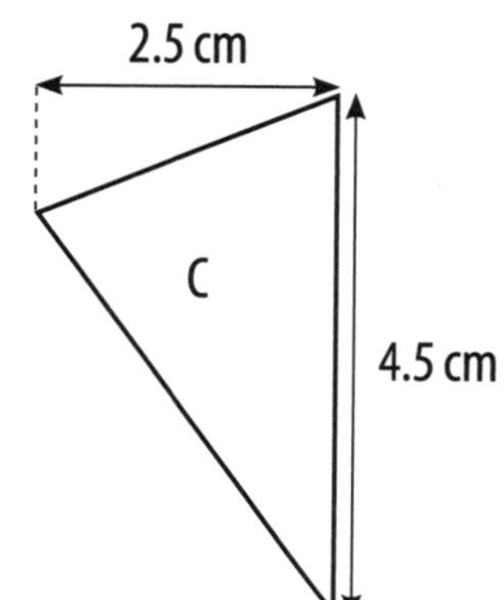


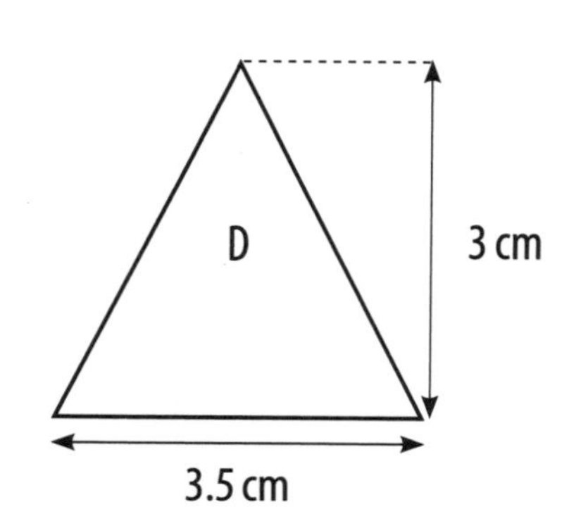


1a (1 mark)

b) Which two triangles have the **same** area?  and

1b (1 mark)

**2** Draw a parallelogram with an area of **15 cm²** on this centimetre-square grid.

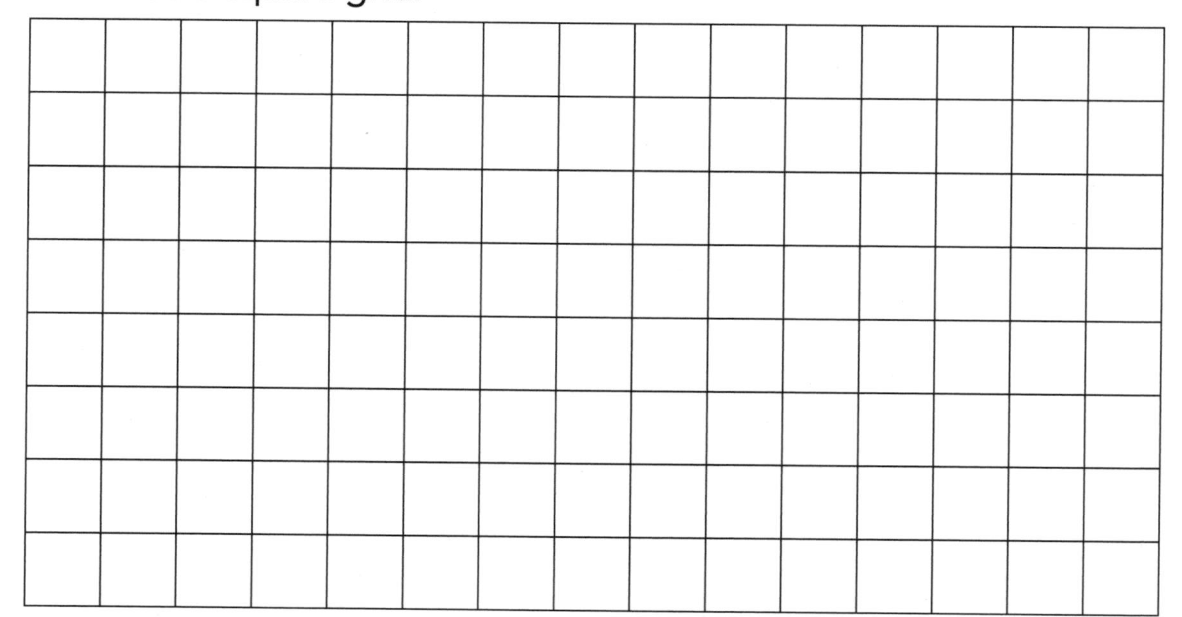

2 (1 mark)

/3 *Total for this page*

**3** Calculate the area of this parallelogram.

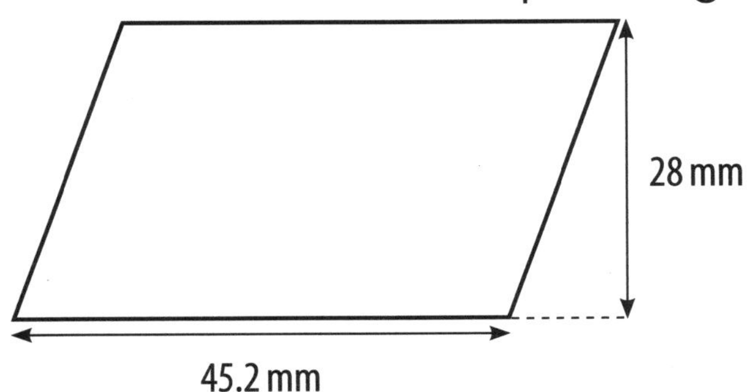


mm²

(1 mark)

**4** Here is a flag.

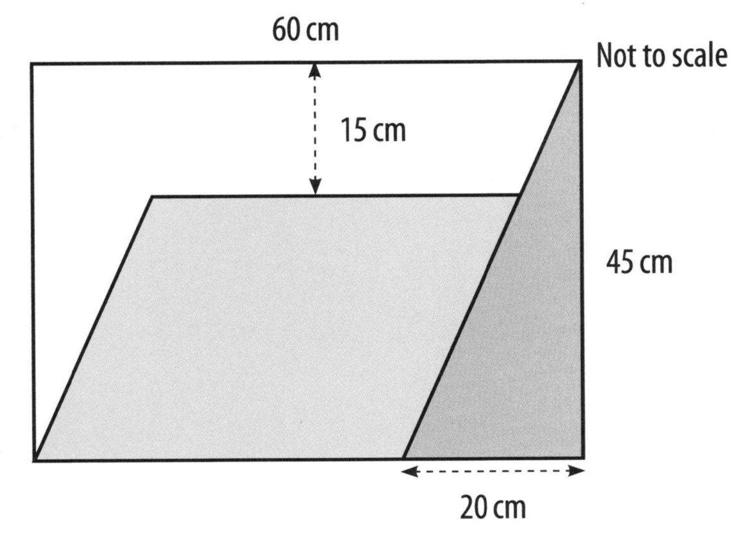


Calculate the area of the white part of the flag. cm²

(2 marks)

**5** The height of a triangle is double the length of its base.

All measurements are in whole centimetres.

a) Circle the possible areas of this triangle.

20 cm²   18 cm²   45 cm²   36 cm²   25 cm²

(1 mark)

b) What do you notice about all the possible areas?

(1 mark)

**Top tip**

- Write any lengths you find on the shape so that you remember to include them in your calculation.

/5 Total for this page

# Volume

To achieve the higher score, you need to:
- calculate, estimate and compare **volumes** of **cubes** and **cuboids** using standard units.

**1** Milly has some 1 $cm^3$ cubes.

She uses them to make a cuboid 7 cm long, 5 cm wide and 2 cm high.

How many cubes does she use? [ ]

(1 mark)

**2** Calculate the volume of this cuboid.

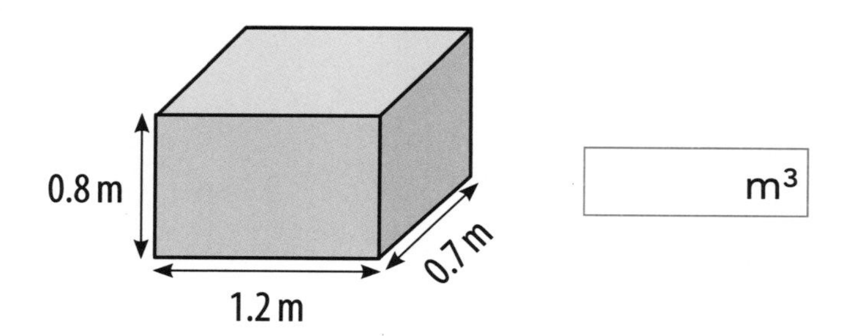


[ ] $m^3$

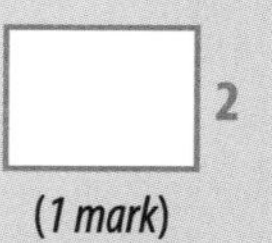
(1 mark)

**3** The volume of a cube is 216 $cm^3$.

What is the length of each edge? [ ] cm

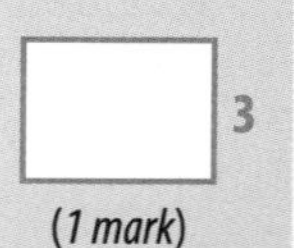
(1 mark)

**4** Jenna uses the formula *Volume = length × width × height* to calculate the volumes of different cuboids.

Draw lines to match the dimensions of the cuboids with their volume. Fill in the missing boxes.

| | |
|---|---|
| 8 cm × 6 cm × 2.5 cm | 135 $cm^3$ |
| 5 cm × [ ] × 2.5 cm | 105 $cm^3$ |
| 4.5 cm × 3 cm × 10 cm | 112.5 $cm^3$ |
| 7 cm × 6 cm × 2.5 cm | [ ] $cm^3$ |

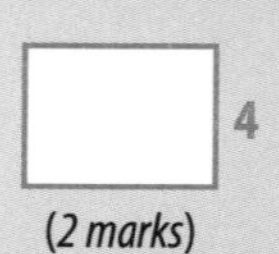
(2 marks)

/5
Total for this page

**5** Jakub uses five of these cuboids to make a tower.

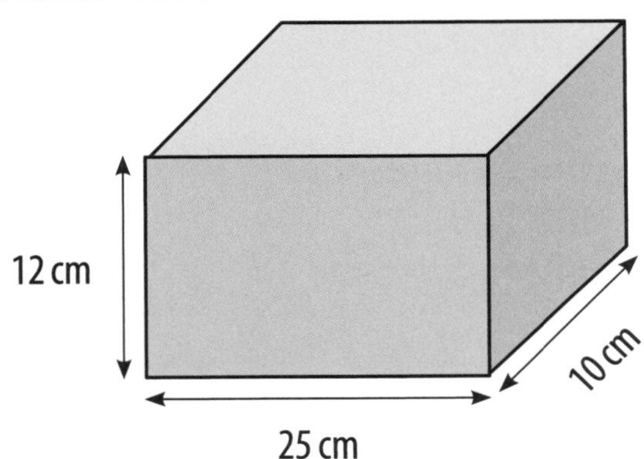


What is the **total** volume of the tower? $\square$ cm³

(1 mark) 5

**6** Anya uses eight cubes to build a tower.

The height of the tower is 32 cm.

What is the volume of **one** cube? $\square$ cm³

(2 marks) 6

**7** The total volume of a cuboid built with 12 cubes is 2,592 cm³.

a) What is the volume of one cube? $\square$ cm³

(1 mark) 7a

b) What is the length of each edge? $\square$ cm

(1 mark) 7b

**8** Pete fills a container with soil.

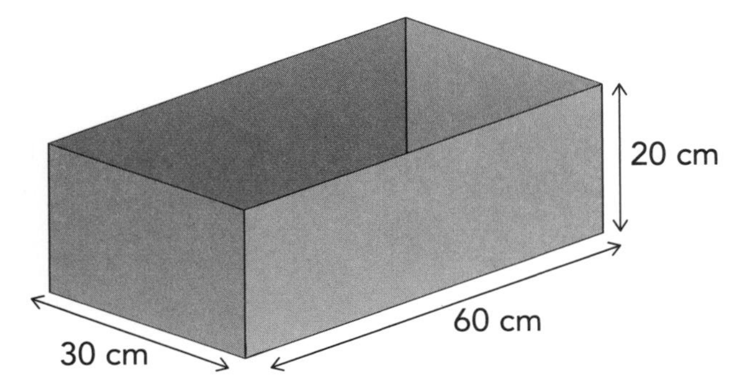

Pete calculates that each bag of soil has a volume of 12,000 cm³.

How much will it cost Pete to fill his container? $\square$

(1 mark) 8

**Top tip**

- When giving the volume of a 3-D shape, make sure all the units are cubed, e.g. cm³ or m³.

/6 Total for this page

# Angles and degrees

To achieve the higher score, you need to:

- ★ recognise **angles** and find missing angles where they meet at a point or are on a straight line or are vertically opposite
- ★ find unknown angles in any triangle, quadrilateral or **regular polygon**.

**1** Five angles of an **equal** size meet at a point.

What is the size of each angle? [ ]°

(1 mark)

**2** Calculate the value of angle $a$.

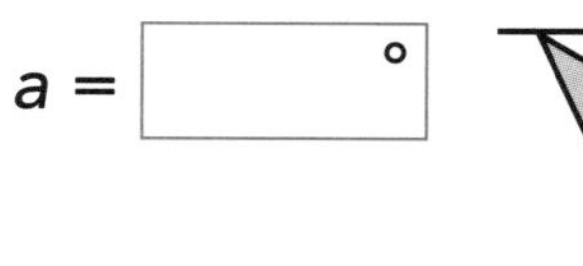

$a =$ [ ]°

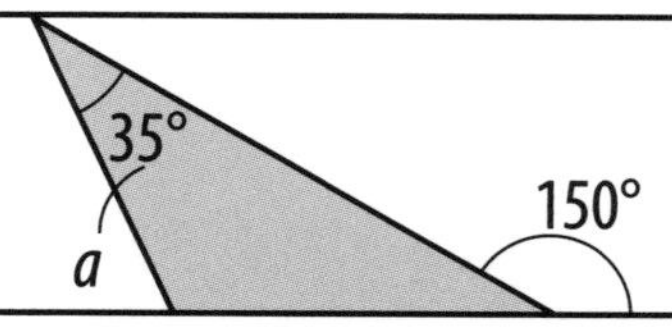


Not to scale

(1 mark)

**3** Calculate angles $a$, $b$ and $c$.

$a =$ [ ]°

$b =$ [ ]°

$c =$ [ ]°

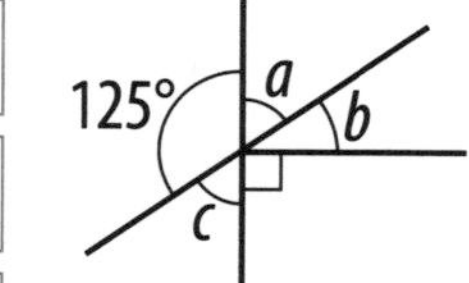


Not to scale

(1 mark)

**4** Angles in a pentagon sum to 540°.

What is the size of **each** angle in a regular pentagon? [ ]°

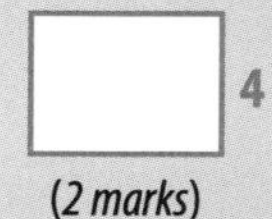

(2 marks)

**5** A regular hexagon is drawn inside a regular trapezium.

Calculate the size of the angles $m$ and $n$.

$m =$ [ ]°

$n =$ [ ]°

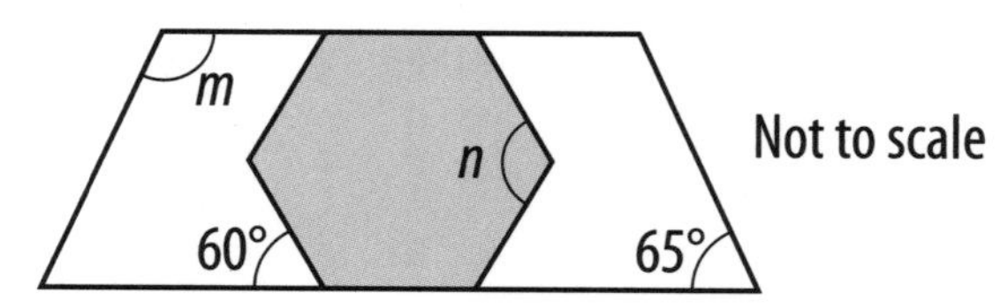


Not to scale

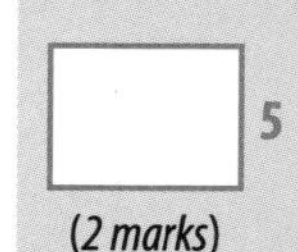

(2 marks)

**Top tip**

- Identify any angles that are of equal size.

/7 Total for this page

# Circles

To achieve the higher score, you need to:
★ illustrate and name parts of circles (including **radius**, **diameter** and **circumference**).

**1** Draw lines to match the parts of a circle.

Complete the missing information on one of the circles.

radius    diameter    circumference

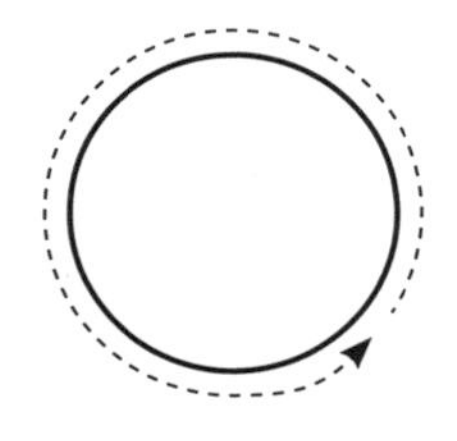 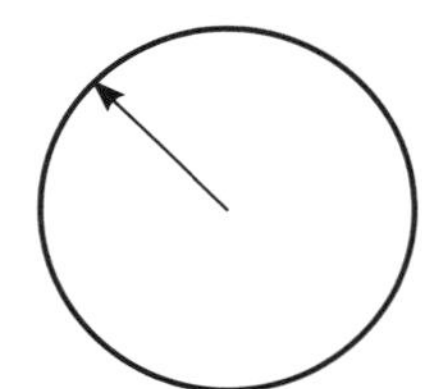 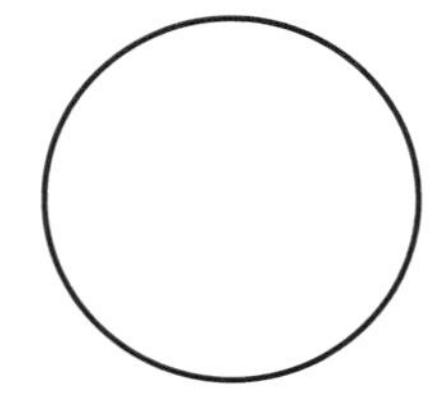

1
(1 mark)

**2** A circle has a diameter of 125 cm.

Circle the length of its radius.

12.5 cm    625 mm    250 cm    62 cm    125 mm

2
(1 mark)

**3** Alfie draws a pattern on squared paper.

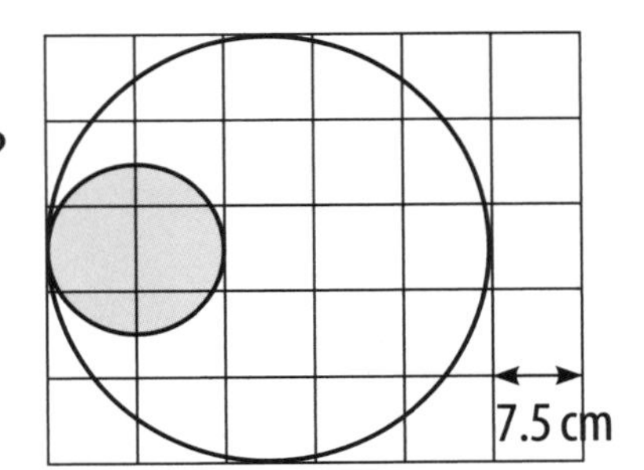


a) What is the diameter of the smaller circle?

[   ] cm

3a
(1 mark)

b) What is the radius of the larger circle?

[   ] cm

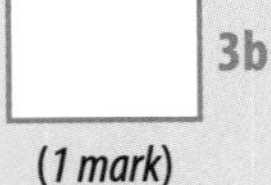
3b
(1 mark)

**4** The radius of one circle is 9 cm.

All the circles are the same size.

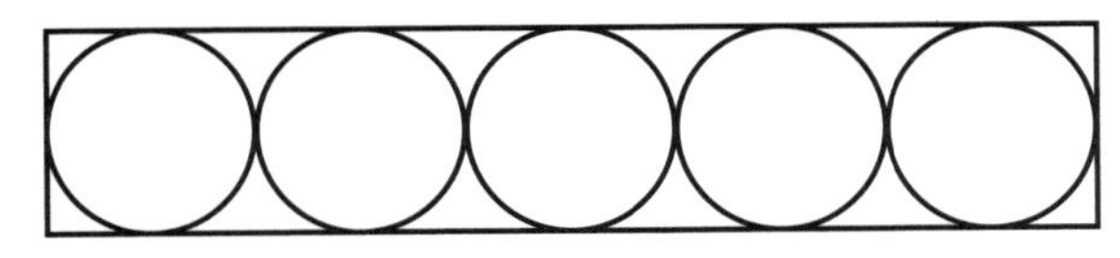

What are the dimensions of the rectangle?

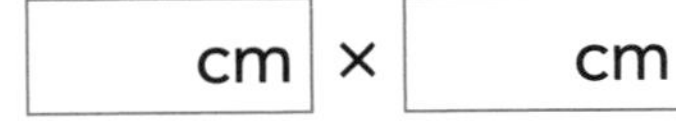
[   ] cm × [   ] cm

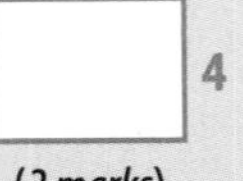
4
(2 marks)

/6
Total for this page

# Coordinates

To achieve the higher score, you need to:
★ describe positions in all four quadrants on a 2-D coordinate grid
★ plot specified points and draw sides to complete a given polygon.

**1** Write the coordinates of the missing vertex needed to complete the rectangle.

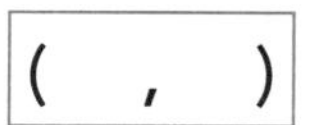

( , )

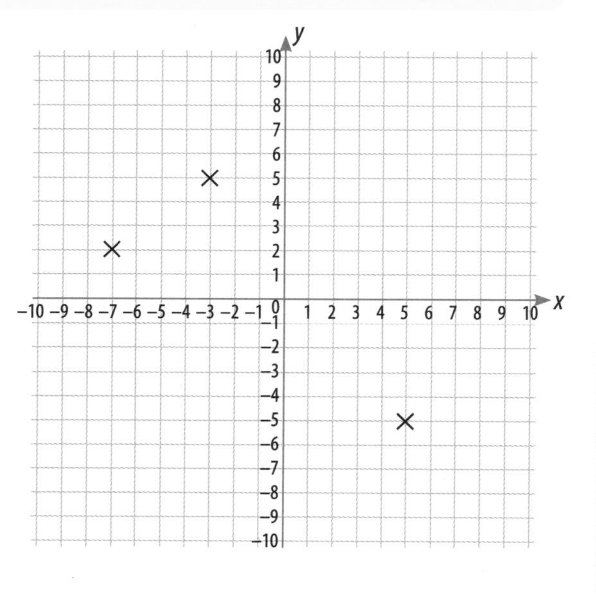

1 (1 mark)

**2** a) Plot these coordinates on the grid.

(1,6) (7,1) (–5,1) (1,–9)

b) Name the shape that has been made.

______________________

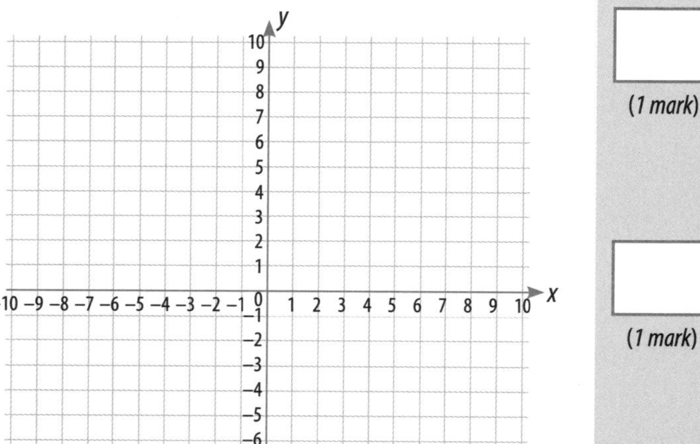

2a (1 mark)

2b (1 mark)

**3** Ruby plots a coordinate at (0,0).

Circle any coordinates Ruby could use to plot the diagonally opposite vertex of a square. The sides of the square should be horizontal and vertical.

(1,–4) (–3,–3) (–5,4) (–4,4) (0,–6)

3 (1 mark)

/4 Total for this page

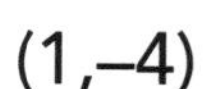

**4** An isosceles triangle has a height of 11 cm.

Fill in the missing coordinates of the vertex.

(1 mark)

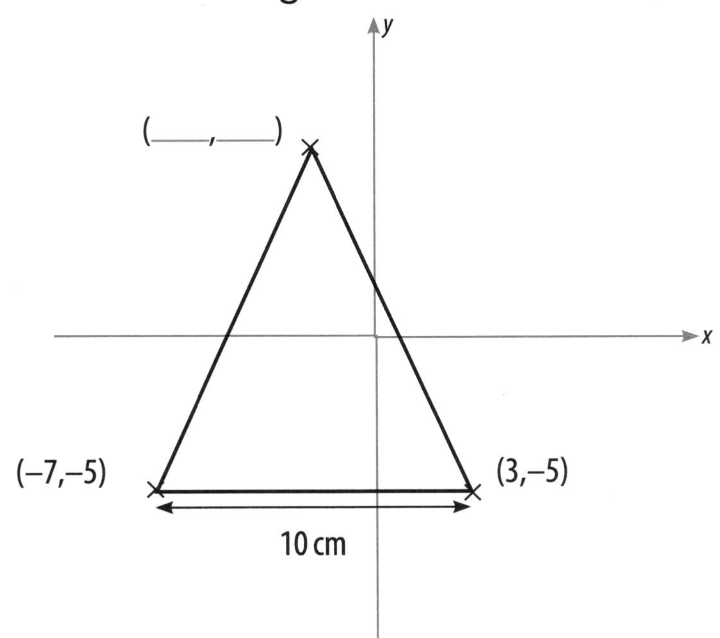


**5** James plans to plot these coordinates and join them to make a square.

(5,4) (8,4) (9,0) (5,0)

Will these coordinates make a square?

Circle your answer. YES / NO

Explain your answer.

___

___

___

(1 mark)

**6** Gwen plots the 3 times table on a grid using coordinates.

Complete the pattern of coordinates:

(0,0), (1,3), (2,6), (3,9), [ , ], (5,15), [ , ], [ , ]

(1 mark)

**Top tip**

- Remember that you can draw on diagrams to help you.

/3

Total for this page

# Translations

To achieve the higher score, you need to:
★ draw and translate simple shapes on the **coordinate plane**.

**1** The **grey triangle** is translated to the same position as the blue triangle.

Describe this translation. ______________________________

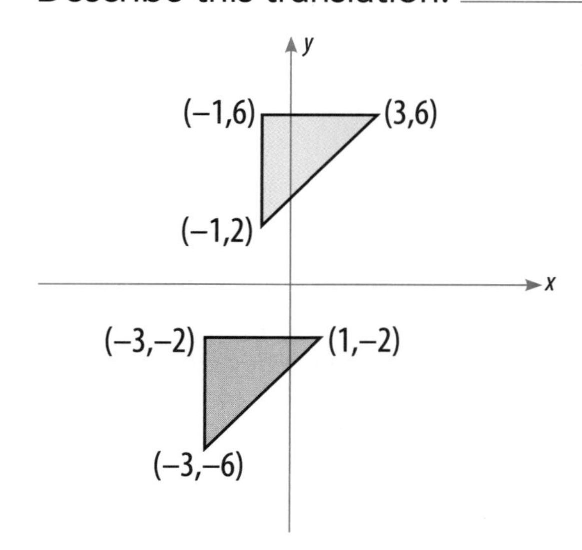

1 (1 mark)

**2** The triangle is translated 3 squares left and 2 squares down.

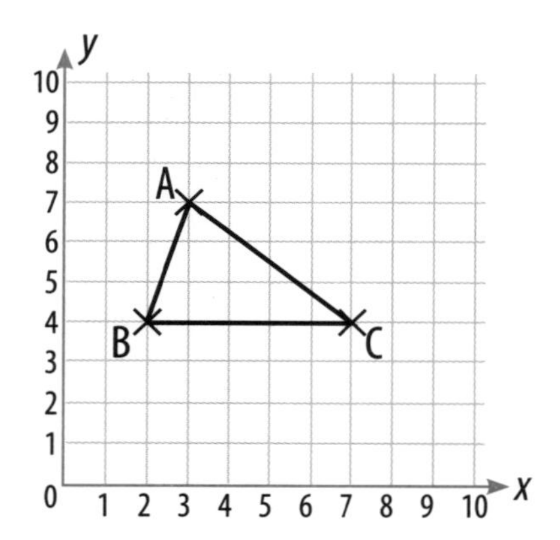


Write the new coordinates of vertex **B**.

( , )

2 (1 mark)

/2

Total for this page

**3** This is the position of a pentagon following a translation.

The original position of vertex **A** was (2,–4).

Describe the translation.

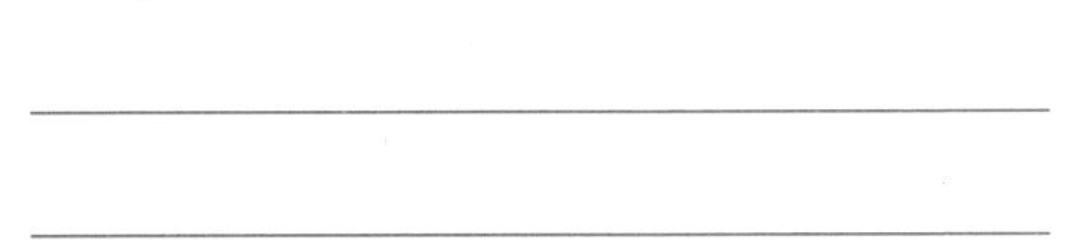

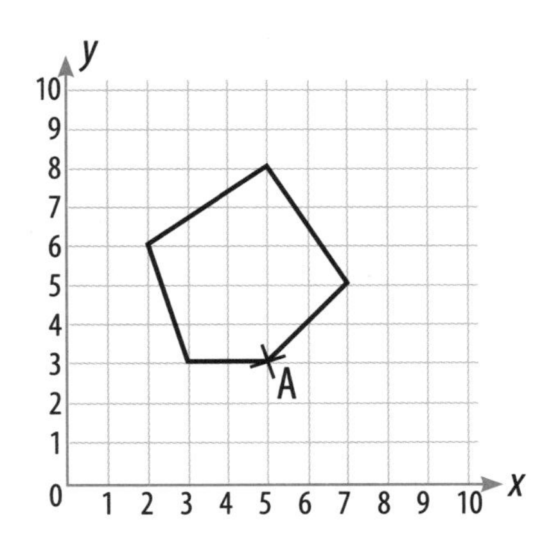


3

(1 mark)

**4** A set of coordinates is moved using the **same** translation each time.

Draw lines below to match the original coordinates with their new positions.

| **Original** | **New** |
|---|---|
| (0,0) | (0,–7) |
| (–2,5) | (3,–4) |
| (–3,–3) | (8,–6) |
| (5,–2) | (1,1) |

Here is a blank coordinate grid to help you work out the translation and match the coordinates.

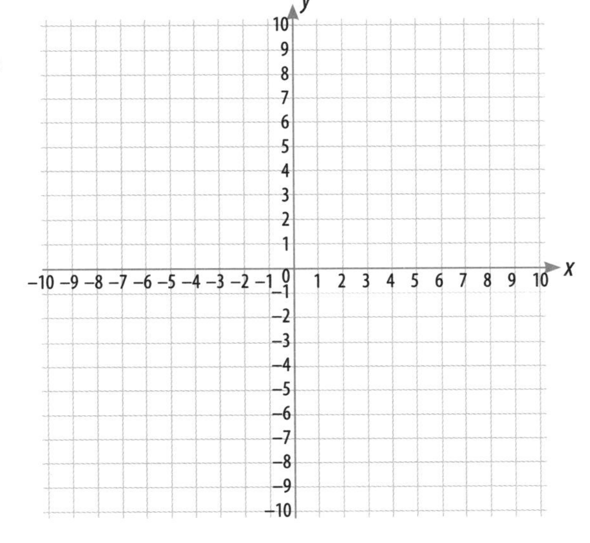


4

(2 marks)

**Top tip**

- Remember you can draw on diagrams to help you.

/3

Total for this page

# Reflections

To achieve the higher score, you need to:

★ reflect shapes in different orientations.

**1** Tick (✓) the shape that is correctly reflected in the mirror line.

(1 mark)

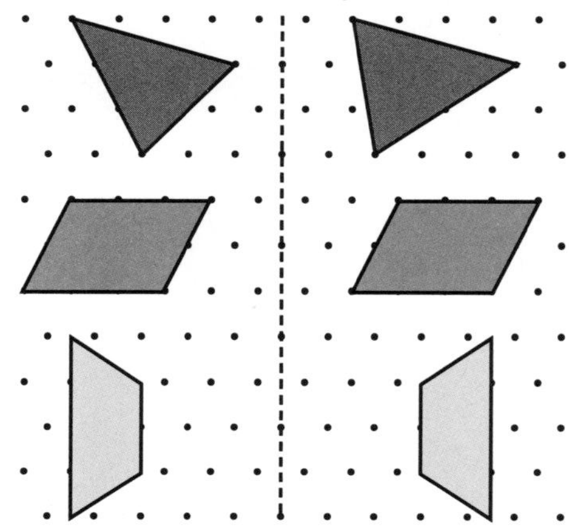

**2** Shade $2\frac{1}{2}$ squares to complete the reflected pattern.

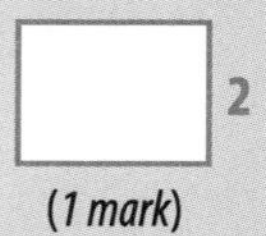

(1 mark)

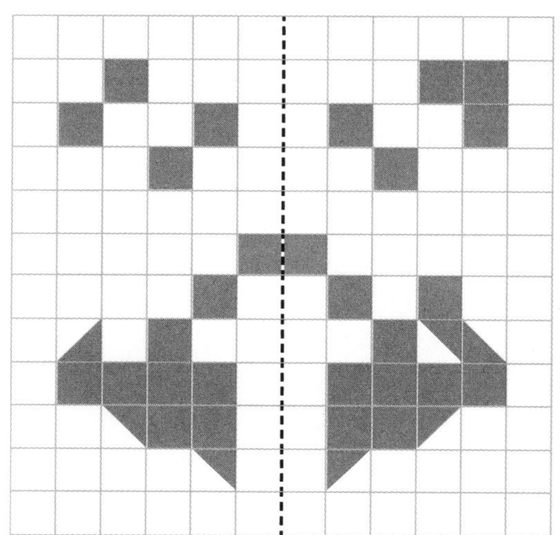

**3** Draw the reflection of the shape in this mirror line.

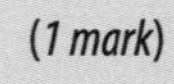

(1 mark)

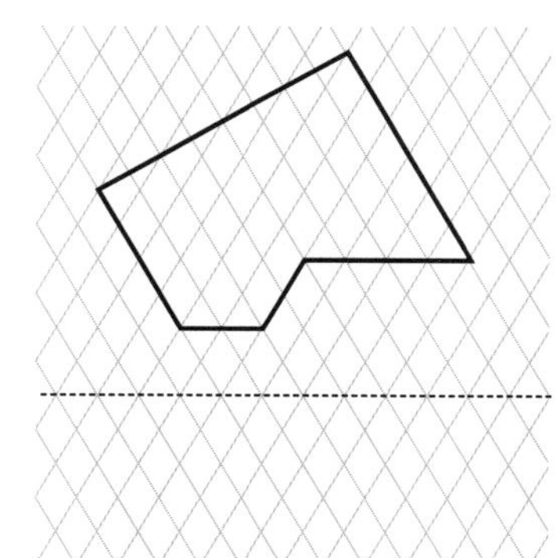

/3

Total for this page

**4** Complete the reflected shape.

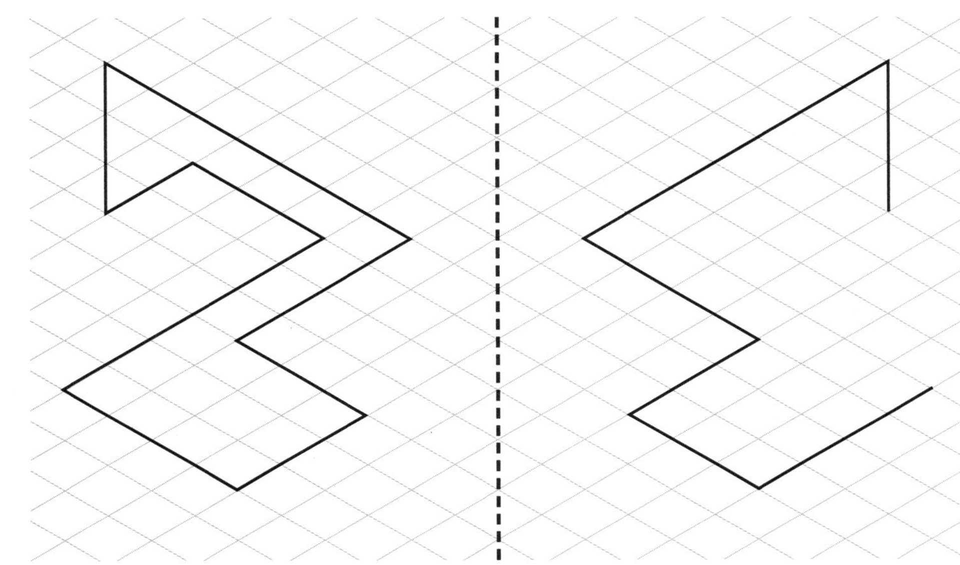

4

(1 mark)

**5** The triangle is translated 5 squares right and 2 squares up on the grid.

It is then reflected in the $x$ axis.

Draw the new position of the triangle.

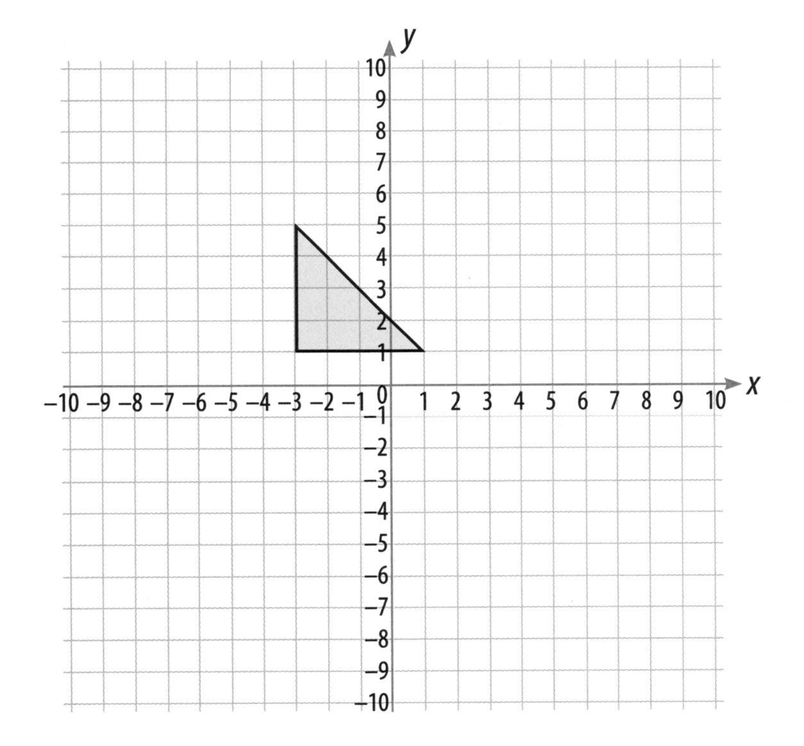


5

(1 mark)

**Top tip**

- Remember that shapes do not change size when they are reflected.

/2

Total for this page

# Tables

To achieve the higher score, you need to:
★ complete missing data in **tables** using problem-solving skills.

**1** This table shows how much money the children save.

| | Sally | Ashott | Sam | Paulo |
|---|---|---|---|---|
| Money saved | £7.75 | £10.99 | £9.20 | £11.49 |

Tick (✓) the statements that are **true**.

Sam needs to save 20% more to have the same amount as Ashott. ☐

Sally has saved just over 75% of £10 ☐

The difference between the amounts saved by two children is $\frac{5}{10}$ of a pound (£). ☐

1 (1 mark)

**2** During Book Week, the children in a school compare their favourite types of books.

| | Story books | Information books | Picture books | Total |
|---|---|---|---|---|
| Boys | $\frac{3}{8}$ | | 25% | 128 |
| Girls | | 49 | 15% | 120 |

a) Complete the missing information in the table.

b) What fraction of the whole school does **not** choose story books? ☐

2a (1 mark)

2b (1 mark)

**3** This table shows the money a town spends on repairs each year.

| | Building repairs | Parks | Road repairs |
|---|---|---|---|
| 2015 | £23,300 | | £18,273 |
| 2016 | £27,308 | £19,826 | |
| 2017 | | £23,785 | £36,208 |
| **TOTAL** | **£70,000** | | **£83,460** |

The total amount spent on parks is $\frac{4}{5}$ of the money spent on buildings.

Complete the missing information in the table.

3 (2 marks)

/5 Total for this page

# Timetables

To achieve the higher score, you need to:
★ complete, read and interpret information in **timetables**.

**1** Tram 3 leaves Stop A exactly 53 minutes later than Tram 2.

Tram 3 takes the same time to travel between stops as Tram 2.

| | Stop A | Stop B | Stop C | Stop D |
|---|---|---|---|---|
| Tram 1 | 11:55 | 12:29 | 13:05 | 13:50 |
| Tram 2 | 12:10 | 12:44 | 13:20 | 14:12 |
| Tram 3 | | | | |

Complete the timetable for Tram 3.

1 (2 marks)

**2** This timetable shows the times that buses stop at different places in the town.

In the evening, they go every 30 minutes after the 16:18 bus.

| Bus garage | **10:18** | **11:18** | **12:18** |
|---|---|---|---|
| Post office | 10:46 | 11:46 | 12:46 |
| School | 11:15 | 12:15 | 13:15 |
| Station | 12:08 | 13:08 | 14:08 |

a) Gemma arrives at the bus garage at 19:23

How long does she have to wait for the next bus?

2a (1 mark)

b) What time does she arrive at the station?

2b (1 mark)

**3** This table shows the departure information at an airport.

| Departure time | Destination | Status |
|---|---|---|
| 11:55 | Berlin | On time |
| 12:10 | Athens | Delayed until 15:20 |
| 12:55 | Stockholm | On time |
| 13:40 | Moscow | Delayed until 16:35 |

| Flight times | | | |
|---|---|---|---|
| Berlin | Athens | Stockholm | Moscow |
| 1 hr 55 | 3 hr 45 | 2 hr 25 | 4 hr 05 |

At what time will passengers arrive in these cities? Berlin

Athens Stockholm Moscow

3 (2 marks)

**Top tip**

- Remember that the 24-hour clock is used for timetables.

/6
Total for this page

# Pictograms

To achieve the higher score, you need to:

★ interpret **pictograms** using a range of skills and strategies.

**1** The pictogram compares the mass of cherries grown by four fruit farms.

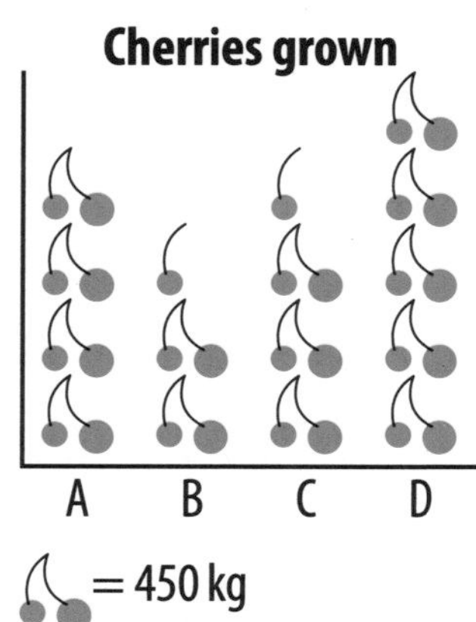


a) What is the **total** mass of cherries grown by the four farms? ______ kg

(1 mark) 1a

b) What **fraction** of the total mass of cherries is grown by farm D? ______

(1 mark) 1b

**2** The pictogram shows the number of spaces available at four car parks.

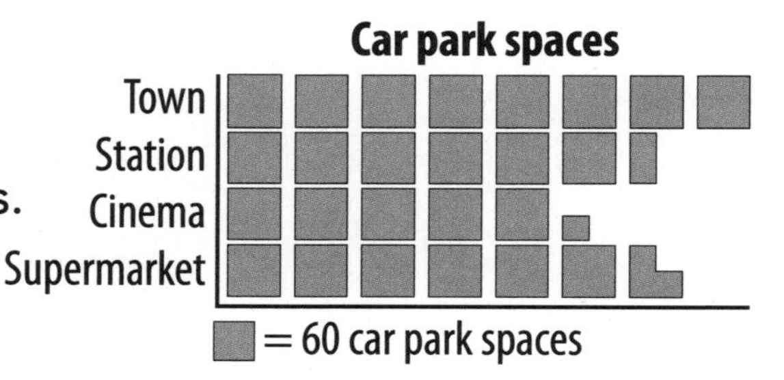


Tick (✓) the statements that are **true**.

☐ The supermarket has 90 more spaces than the cinema.

☐ All car parks have more than 360 spaces.

☐ The ratio of town spaces to station spaces is 16:13

(1 mark) 2

**3** The pictogram shows the number of trees planted in rainforests by a world charity.

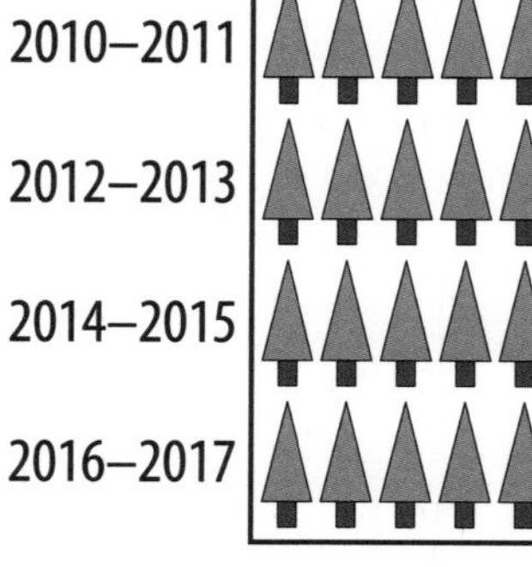


a) How many **more** trees were planted in 2014–2015 than in 2010–2011? ______

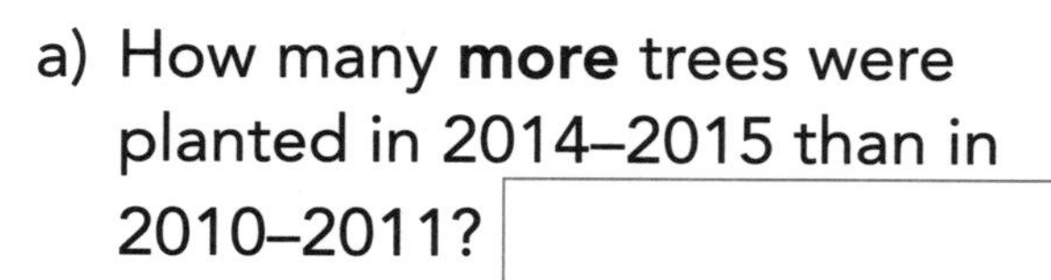

(1 mark) 3a

b) How many trees in **total** were planted between 2012 and 2015? ______

(1 mark) 3b

c) What **percentage** of the total number of trees was planted in 2016–2017? ______ %

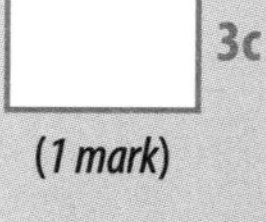

(1 mark) 3c

**Top tip**

- Remember to check the scale each time.

/6 Total for this page

# Bar charts

To achieve the higher score, you need to:

★ interpret **bar charts** using a range of skills and strategies.

**1** This bar chart compares the percentage of a computer game completed by each player.

| Peter | Remi | Amber | Jayden | Kia |
|---|---|---|---|---|
| | 65% | | | 80% |

a) Complete the table.

b) Draw the bar for Remi on the chart.

**Computer game completed**

Peter

Remi

Amber

Jayden

Kia

Percentage (%) of computer game completed

1a (1 mark)

1b (1 mark)

**2** Natasha thinks that the mass of the pebbles is $\frac{3}{8}$ kg **lighter** than the mass of the soil.

Is she correct?

Circle your answer. YES / NO

Explain your answer.

______________________________

______________________________

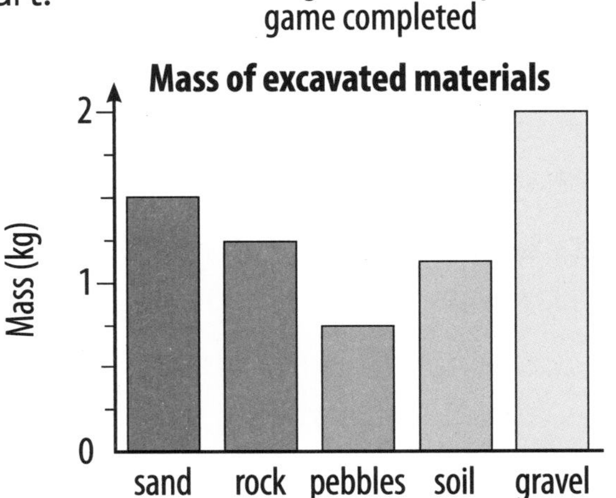


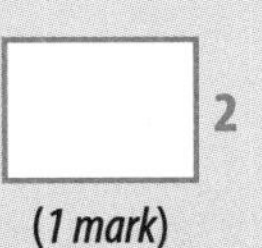


**3** The bar chart compares the profits of five companies from 2015 to 2017.

a) Which companies made more than £50,000 profit in 2016?

______________________________

b) In which year did company B make its greatest profit?

c) What is the difference between the total profits made by companies C and E?

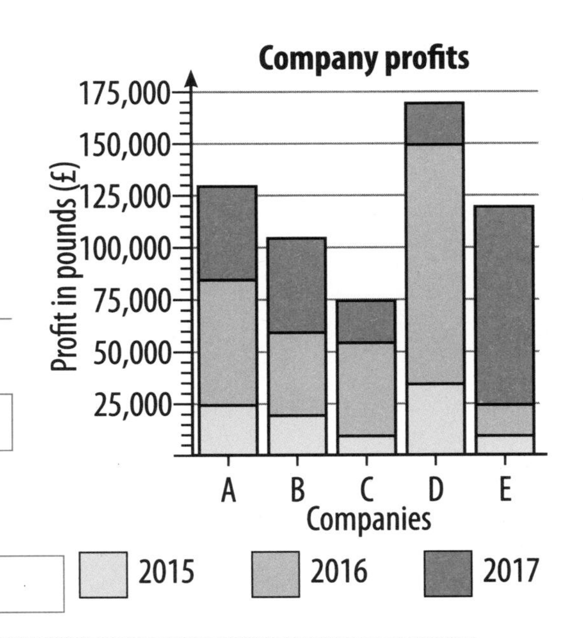


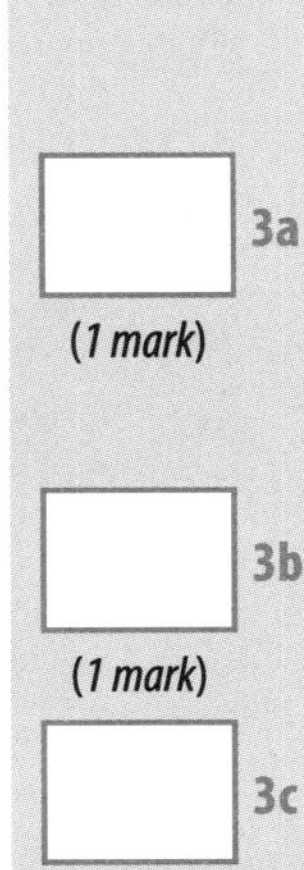


**Top tip**

- Work out the scale first and then write the numbers on the chart to help you.

/6

*Total for this page*

# Pie charts

To achieve the higher score, you need to:
- ★ interpret and construct **pie charts**
- ★ use pie charts to solve problems.

**1** A class takes part in four different activities to raise money for charity.

The table shows how much money each activity raised.

| Sponsored walk | £30 |
|---|---|
| Cake sale | £20 |
| Own clothes day | £30 |
| Fun run | £40 |
| **TOTAL** | **£120** |

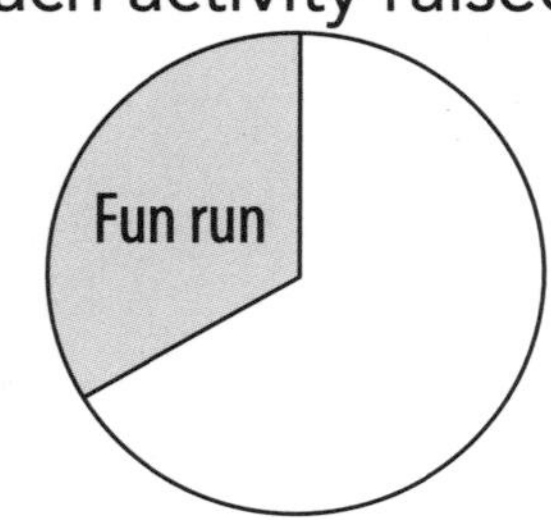


Use the information to complete the pie chart.

1 (2 marks)

**2** This pie chart compares the masses of fruit sold at a supermarket during one weekend.

apples
pears
strawberries
oranges
blueberries
bananas

The supermarket sold an equal mass of apples and bananas, together totalling 520 kg.

It also sold an equal amount of apples, pears, strawberries and blueberries.

a) How many kilograms of oranges were sold? ______ kg

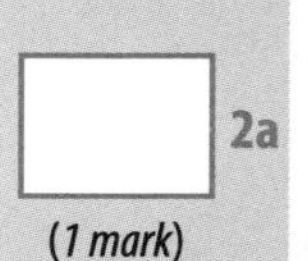

2a (1 mark)

b) What was the **total** mass of fruit sold?

______ kg

2b (1 mark)

**3** A travel agency uses a pie chart to compare the number of holidays booked to different destinations.

UK
France
Spain
Greece

800 holidays were booked in the **UK**.

$\frac{1}{8}$ of the holidays were to **Greece**.

a) How many holidays were booked in **total**? ______

3a (1 mark)

b) How many holidays were booked to **Greece**? ______

3b (1 mark)

c) The number of bookings to **France** and **Spain** are **equal**.

How many bookings were made to **France**? ______

3c (1 mark)

**Top tip**
- Remember to label each part of a pie chart when asked to complete it.

/7

*Total for this page*

# Line graphs

To achieve the higher score, you need to:
★ interpret and construct line graphs and use these to solve problems.

**1** This line graph shows the route taken in a mountain bike race.

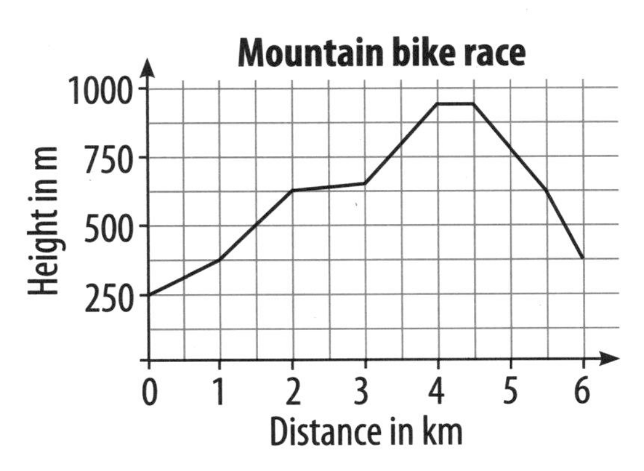


a) Complete the table using the line graph to help you.

| Distance (km) | 1 | | | 4 | 5.5 |
|---|---|---|---|---|---|
| Height (m) | | 625 | 875 | | |

(1 mark) 1a

b) How many kilometres of the route were **below** a height of **625 m**? ______ km

(1 mark) 1b

**2** This line graph shows the number of viewers at two different times of the day over a number of months.

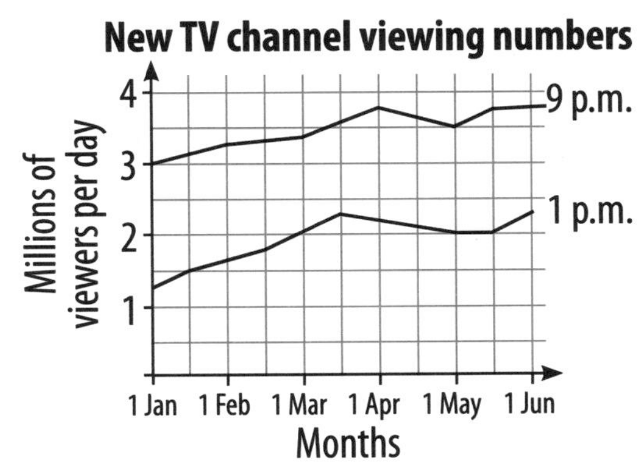


a) How many viewers at 9 p.m. are shown for 1 June?

(1 mark) 2a

b) Approximately how many more viewers were there at 9 p.m. than 1 p.m. for 1 March?

(1 mark) 2b

c) Explain why you think the viewing numbers are lower at 1 p.m. than at 9 p.m.

(1 mark) 2c

## Top tip

- Remember to read the graph carefully to work out the value of unmarked divisions and also to estimate the value of spaces between the lines.

/5 Total for this page

# Averages

To achieve the higher score, you need to:
★ calculate the **mean** as an **average** for simple sets of discrete data in different contexts.

**1** This table shows the heights of four friends.

Calculate the mean height.

| Kamal | Nita | Ella | Jamie |
|---|---|---|---|
| 1.85 m | 1.6 m | 1.75 m | 1.8 m |

[ ] m

*(1 mark)* 1

**2** The mean height of the five friends is 174 cm.

| Kamal | Nita | Ella | Jamie | Sean |
|---|---|---|---|---|
| 1.85 m | 1.6 m | 1.75 m | 1.8 m | |

How tall is Sean? [ ] m

*(1 mark)* 2

**3** The mean of three mystery capacities is 635 ml.

Circle the **three** mystery capacities.

765 ml     475 ml     570 ml     665 ml     835 ml

*(1 mark)* 3

**4** Find the mean of this set of numbers.

[ ]

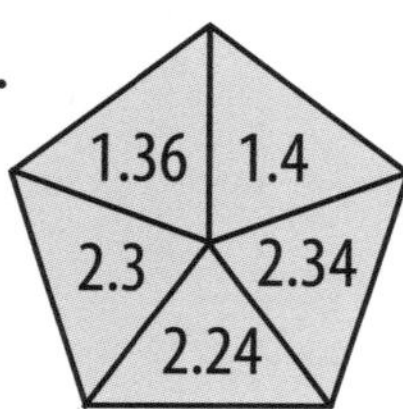


*(1 mark)* 4

**5** This bar chart shows the number of points scored by three of the teams in a quiz.

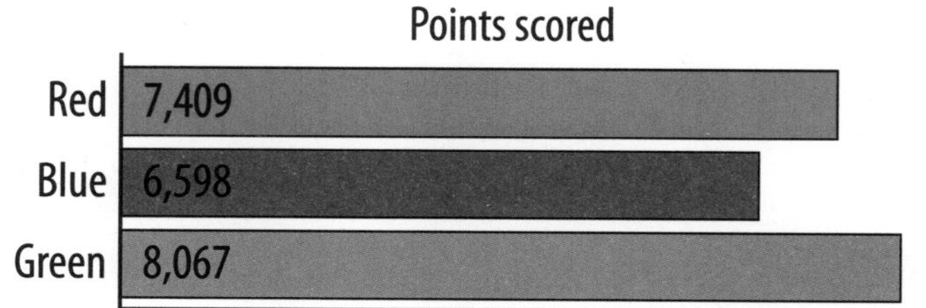


Yellow team also took part in the quiz.

The **mean** number of points scored by all four teams is 7,365.

Calculate the number of points scored by the yellow team. [ ]

*(1 mark)* 5

## Top tip

- The mean value is **not** one of the data items. It is calculated as the sum of all the values divided by the number of values.

/5 *Total for this page*

# Answers

## Number and place value

### Place value (page 6)

1 60 thousand (600,000), 6 hundredths (0.06), 6 thousandths (0.006)
2 2,132,000 and 6,368,000 circled
3 'The digit 1 in 2,107,225 has the value ten thousand.' ticked
4 1,340; 1,350; 1,430; 1,450; 1,530; or 1,540
5 4,960,005

### Roman numerals (page 7)

1 383
2 CXLV
3 CLVII and CXLVIII circled
4 a) MDCLXVI
b) Award **2 marks** for an answer that refers to all seven Roman numerals being used **and** that they are written in order from greatest value to least value.
Award **1 mark** for either of the above points.
5 Award **1 mark** for correct answer NO and an explanation to show that either:
- the film was made in 1967 so Gran must have been born in 1958
- the film was made only 9 years after the year that Gran was born
- 1957 is 10 years before the film was made.

**Do not award** any marks for a NO response without a valid explanation, or simply an explanation that Gran was born later than 1957 as this does not imply that 1967 was identified.

## Number – Addition, subtraction, multiplication and division

### Addition and subtraction (page 8)

1 9,729; 12,197
2 4,471
3

| | | | | | |
|---|---|---|---|---|---|
| | 5 | 7 | 2 | **7** | 3 |
| – | 2 | **3** | 7 | 4 | **4** |
| | 3 | 3 | 5 | 2 | 9 |

4 42,270
5 a) 52
b) Award **2 marks** for correct answer 56.
If the answer is incorrect, award **1 mark** for evidence of appropriate working, e.g.:
• 98 – 23 – 19 = wrong answer
OR
• 23 + 19 = 42
• 98 – 42 = wrong answer

### Squares and cubes (page 9)

1 9
2 $8^2$; $2^3$.
3 8 × 8 × 8
4 triangle = 216, pentagon = 121
5 Award **2 marks** for correct answer 9 × 5 (or vice versa).
Award **1 mark** for an error resulting in incorrect area of base, as long as dimensions follow through, e.g. area incorrectly calculated as 40 cm² and dimensions given as 4 × 10.

### Common multiples (page 10)

1 168 or 192
2 72
3 360
4 909 and 480 circled
5 Award **3 marks** for correct answer £10.10.
Award **2 marks** for identifying 840 mm and one arithmetic error when calculating change.
Award **1 mark** for identifying 840 mm and correct calculating of £9.90 as the cost of two pieces, but not completing the last step of finding the amount of change.

### Common factors (page 11)

1 1; 2; 4; 8
2 9 and 12

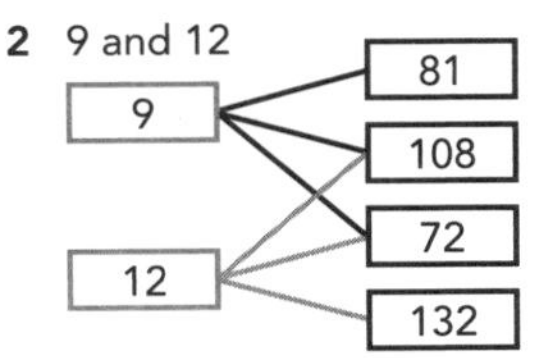


3 3; 4; 6
4 4; 8; and 6 ticked
5 10; 20; 40; 70; 140; 280

### Prime numbers and prime factors (page 12)

1 61; 67; and 71 circled
2 210
3

| | | | |
|---|---|---|---|
| 3 | 5 | 2 | 5 |
| 5 | 2 | 2 | 2 |
| 2 | 5 | 5 | 3 |

or

| | | | |
|---|---|---|---|
| 3 | 5 | 2 | 5 |
| 5 | 2 | 2 | 2 |
| 2 | 3 | 5 | 5 |

4 36 and 144
5 35; 28; and 21

### Multiplying and dividing by 10, 100 and 1,000 (page 13)

1 a) 1,000 b) 100
2

13.6
14
23
23.4
22.4

3 4.75
4 0.325
5

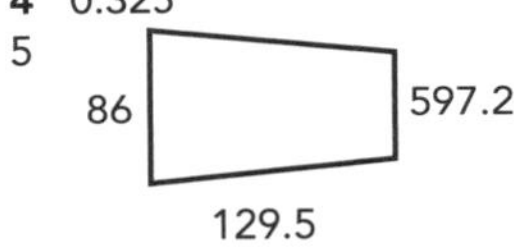


### Long multiplication (page 14)

1 Award **2 marks** for correct answer £3,024.
Award **1 mark** for a correct method.
2

| | | | | | |
|---|---|---|---|---|---|
| | | | **4** | 6 | 3 |
| × | | | | **7** | 8 |
| | | 3 | 7 | 0 | 4 |
| | 3 | 2 | 4 | 1 | 0 |
| | 3 | 6 | 1 | 1 | 4 |

3 6,300
4 Award **2 marks** for correct answer 34,290.
Award **1 mark** for a correct method.
5 15; 33; 24

### Long division (page 15)

1 309 boxes
2 123.5
3 588
4 a) £432.66 b) £0.10 or 10p
5 $\frac{2}{3}$ and 8 circled
6 £398.50

### Correspondence (page 16)

1 24
2 12
3 a) 12 b) 15

### Order of operations (page 17)

1 152
2 95

3
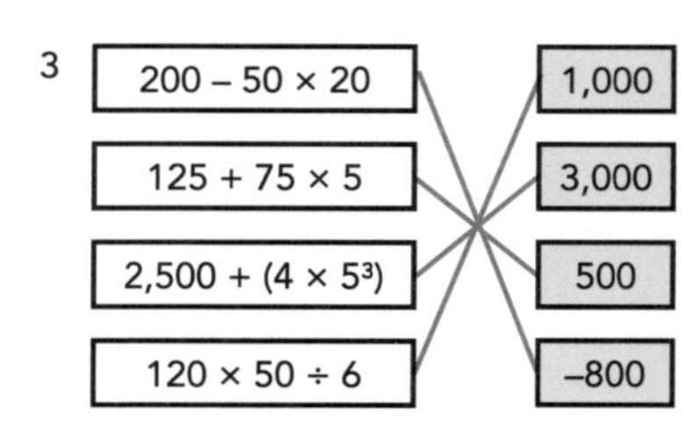


4 16 + $6^2$ = $10^2$ – 4 × 12

5 Award **1 mark** for correct answer NO and an explanation to show that either:
- the order of operations means that multiplication and division in a calculation can be carried out in any order as they have the same importance or level of priority, but addition has less importance and should be carried out after division or multiplication
- reference to BIDMAS or similar to explain that addition must be carried out after division
- worked-through examples using order of operations proving Jayden to be correct but Anna incorrect.

**Do not award** any marks for simply stating 'because of the order of operations' as this does not imply an understanding of the priorities.

## Number – Fractions, decimals and percentages

### Ordering fractions (page 18)

1 Award **2 marks** for correct answer >, =, >.
Award **1 mark** for any two correct.
2 triangle ticked
3 $\frac{5}{12}$ $\frac{3}{4}$ $\frac{4}{5}$ $\frac{7}{8}$ $1\frac{1}{2}$
4 A, C, B

### Adding and subtracting fractions (page 19)

1 $\frac{7}{8}$
2 a) $1\frac{4}{9}$ or $\frac{13}{9}$ b) $\frac{5}{9}$
3 $1\frac{23}{24}$
4 $1\frac{5}{18}$ or $\frac{23}{18}$ (or accept $1\frac{10}{36}$ or $\frac{46}{36}$ if original fraction $\frac{8}{12}$ not simplified to $\frac{2}{3}$)
5 Award **2 marks** for correct answer $1\frac{7}{8}$.
Award **1 mark** for the correct answer that has not been simplified $1\frac{21}{24}$ or $\frac{45}{24}$.

### Multiplying fractions (page 20)

1 $3\frac{3}{5}$ (accept $\frac{18}{5}$)
2 $\frac{14}{25}$
3 $1\frac{7}{8}$
4 $\frac{1}{2}$
5 $1\frac{1}{2}$
6 $\frac{9}{80}$

### Dividing fractions (page 21)

1 $\frac{5}{24}$
2 5
3 $\frac{1}{15}$
4 Award **2 marks** for correct answer $\frac{1}{16}$.
Award **1 mark** for sight of the calculation $\frac{3}{8} \div 6$ or $\frac{3}{8} \times \frac{1}{6}$ if the final answer is incorrect.
5 $\frac{3}{4}$m

### Changing fractions to decimals (page 22)

1 4.75
2 0.625
3 3.67
4 8.75 and $8\frac{3}{4}$ circled
5 $\frac{2}{5}$ and 0.4
6 $\frac{3}{8}$; 0.375

### Rounding decimals (page 23)

1 a) 36.5 b) 36.47
2 63.89
3 Bar drawn at 1.5 million
4 2.77 kg, 2.849 kg and 2.809 kg circled
5 3.0483 litres and 3.0349 litres circled

### Adding and subtracting decimals (page 24)

1 24.447
2 a) 1.4 + 2.24 b) 2.34 – 2.3
3 a) 2.382 b) 4.418
4 14.5
5 Award **2 marks** for correct answer 0.585.
Award **1 mark** for an appropriate method with no more than one mathematical error, e.g.
1.675 + 0.24 = 1.915
2.5 – 1.915 = wrong answer

### Multiplying decimals (page 25)

1 0.8; 80; 0.08
2 0.8
3 a) 3.6; 8.1; and 0.56 circled
b) Explanation shows that all the answers must be related facts of the nine times table as Sara is multiplying by a whole number. So 0.56 cannot be an answer because 56 is not a multiple of 9. Although 36 and 81 are multiples of 9, Sara only multiplies by a single digit and she would need to multiply 0.09 by 40 to get an answer of 3.6, etc.
4 Award **2 marks** for correct answer 0.21; 5.6; 0.49; 0.09.
Award **1 mark** for any three correct.
5 0.027

### Percentages (page 26)

1 Freddy; £15
2 Award **2 marks** for correct answer £640.
Award **1 mark** for a correct method.
3 Award **2 marks** for correct answer 195.
Award **1 mark** for a correct method.
4 Both ticked for the mark.

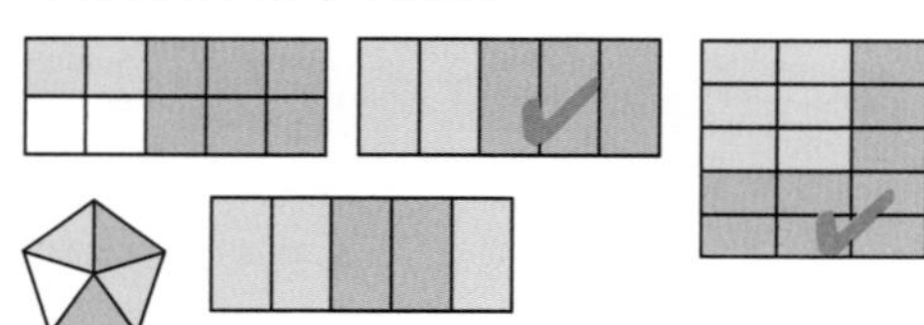

5 Award **1 mark** for correct answer YES and an explanation that shows either:
- 60% is the same as $\frac{6}{10}$ and an equivalent fraction of $\frac{6}{10}$ is $\frac{24}{40}$
- $\frac{24}{40}$ can be simplified to $\frac{6}{10}$, which is equivalent to 60%.

**Do not award** any marks for simply restating that $\frac{24}{40}$ is equivalent to 60%.

## Ratio and proportion

### Ratio (page 27)

1 60
2 84 cm circled
3 49
4 Award **2 marks** for all correct.
Award **1 mark** for any four correct.

| Rectangular tiles | 18 | 54 | 66 | 120 | 180 |
|---|---|---|---|---|---|
| Circular tiles | 15 | 45 | 55 | 100 | 150 |

### Proportion (page 28)

1 $\frac{5}{8}$ or $\frac{20}{32}$
2 525
3 £3.60
4 30
5 675
6 Award **2 marks** for correct answer 4.2
If the answer is incorrect, award **1 mark** for evidence of an appropriate method, e.g.
3.5 × 6 = 21
21 ÷ 5 = wrong answer

### Scaling problems (page 29)

1 Accurate drawing of a square with sides of 3 cm (accept 1 mm margin of error).
2 12.8 and 14.4
3 120
4 11.7

### Unequal sharing (page 30)

1 a) 150
b) 210
2 Mo 60, Darren 20, Belle 40
3 a) 100 £2 coins and 180 £1 coins
b) £380
c) £82
Award **2 marks** for correct answer.
Award **1 mark** for a correct method.

## Algebra

### Algebra (page 31)

1 $(n \div 4) + 5$ circled
2 61; 21; 13.8
3 a) $3a + b$ (or vice versa) b) $4b + a$ (or vice versa)
4 $y = x + 3$

### Using formulae (page 32)

1 a) 52.4 b) 16.8
2 $b = 360 - 45 - 180$
3 Award **2 marks** for all correct.
Award **1 mark** for any two correct.

| Fahrenheit (°F) | 77 | 50 | 59 |
|---|---|---|---|
| Celsius (°C) | 25 | 10 | 15 |

### Solving equations (page 33)

1 $m = 30$ and $n = 4$
2 □ = 4 ⬠ = 15
3 e.g. 3 m × 8 m or 4 m × 6 m
4 6

### Linear number sequences (page 34)

1 8 14 20 26
2 $4n - 1$ circled
3 a) 35 b) 120
4 r = $12n$ − 6

## Measurement

### Measures (page 35)

1 a) 47 b) 0.325
2 20 and 40 circled
3 Award **2 marks** for correct answer 4.99
Award **1 mark** for 499 as this shows evidence of a correct method and conversion of mm to cm.
4 Award **2 marks** for correct answer 105.94
Award **1 mark** for a correct method.

### Converting metric units (page 36)

1 Award **2 marks** for all correct.
Award **1 mark** for any six correct.

| km | 0.0121 | 0.45 | 0.143 | 0.137 |
|---|---|---|---|---|
| m | 12.1 | 450 | 143 | 137 |
| cm | 1,210 | 45,000 | 14,300 | 13,700 |

2 3.323
3 £5.45
4 Award **1 mark** for correct answer YES and an explanation that shows a worked example giving an answer of 63 hours (3,780 minutes) left, which is just over half of 120 hours (60 hours).
**Do not award** any marks for simply showing that half of 120 hours is 60 hours.

### Metric units and imperial measures (page 37)

1

| centimetres | 225 | 660 | 157.5 |
|---|---|---|---|
| feet | 7.5 | 22 | $5\frac{1}{4}$ |

2 720
3 8.5
4 12
5 24.5

### Perimeter and area (pages 38–39)

1 Perimeter of 16, e.g.

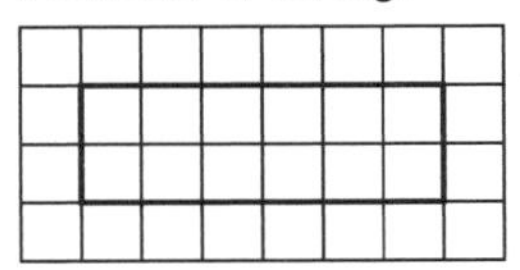

2

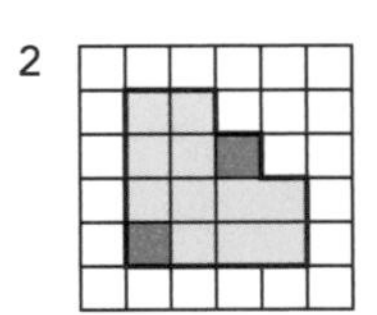

or

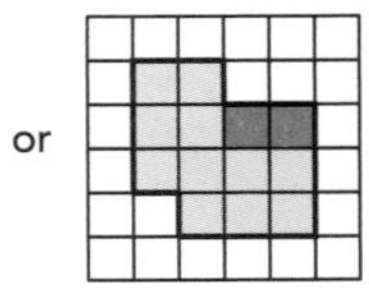

or

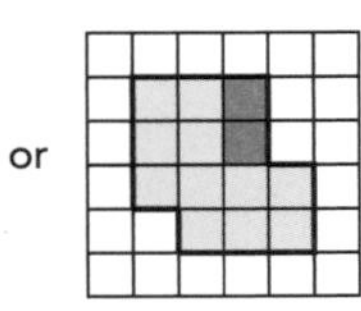

3 Award **1 mark** for shortest 80; award **1 mark** for longest 170 (award **1 mark** for **both** 800 and 1,700 if the answers are given in mm).
4 769
5 Award **2 marks** for correct answer 9 cm.
Award **1 mark** for a correct method.
6 Award **3 marks** for the correct answer of 28
If the answer is incorrect, award **2 marks** for evidence of an appropriate method with no more than one arithmetic error, e.g.
10 × 10 = 100
12 x 6 = 72
100 – 72 = wrong answer
Award **1 mark** for evidence of an appropriate method that has more than one arithmetic error.

### Area of parallelograms and triangles (pages 40–41)

1 a) triangle B ticked b) A and D
2 Any parallelogram with base and height that are factor pairs of 15, e.g.

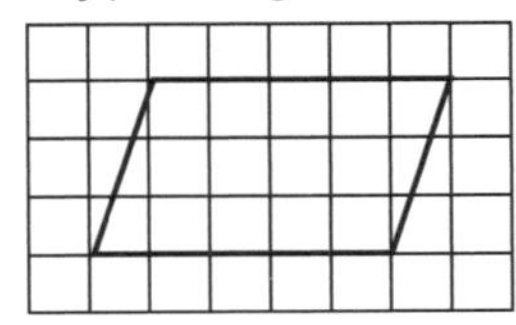

3 1,265.6
4 Award **2 marks** for correct answer 1,050
Award **1 mark** for a correct method.
5 a) 36 cm² and 25 cm² circled
b) They are square numbers.

### Volume (pages 42–43)

1 70
2 0.672
3 6
4 Award **2 marks** for all correct.
Award **1 mark** for all matched correctly but incorrect value for either 9 or 120.

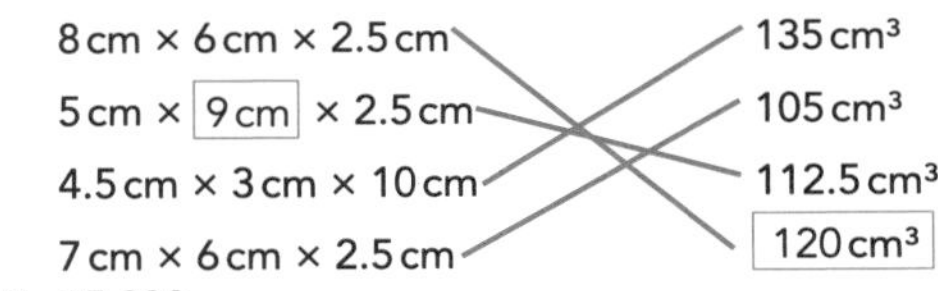


5 15,000
6 Award **2 marks** for correct answer 64
Award **1 mark** for a correct method but an incorrect answer, e.g. 32 cm ÷ 8 = 4 cm, 4 × 4 × 4 = error.
7 a) 216 b) 6
8 £11.97

## Geometry – Properties of shapes

### Angles and degrees (page 44)

1 72
2 65
3 $a = 55$ $b = 35$ $c = 55$
4 108
5 $m = 115$ and $n = 120$

### Circles (page 45)

1 All correctly joined and a diameter added to the circle on the right for the mark.

radius diameter circumference

2 625 mm circled

3 a) 15

b) 18.75

4 Award **2 marks** for correct answer 18 × 90

Award **1 mark** for a correct method but an incorrect answer, e.g. 2 × 9 cm = 18 cm, 18 × 5 = wrong answer.

## Geometry – Position and direction

### Coordinates (pages 46–47)

1 (1,–8)

2 a)

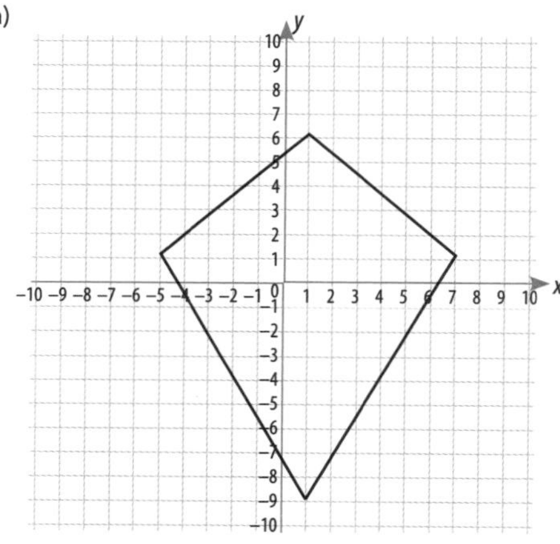

b) kite

3 (–3,–3) and (–4,4) circled

4 (–2,6)

5 Award **1 mark** for correct answer NO and an explanation either:

- because the coordinate (8,4) should be positioned at (9,4)
- (9,0) and (5,0) are four squares apart but (5,4) and (8,4) are not
- all the sides should be four squares but the side at the top will only be three squares.

**Do not award** a mark for simply saying it will not make a square or the shape is not square.

6 4,12; 6,18; 7,21

### Translations (pages 48–49)

1 2 right and 8 up

2 (–1,2)

3 3 right and 7 up

4 Award **2 marks** for all correct.

Award **1 mark** for any two correct.

| **Original** | **New** |
|---|---|
| (0,0) | (0,–7) |
| (–2,5) | (3,–4) |
| (–3,–3) | (8,–6) |
| (5,–2) | (1,1) |

### Reflections (pages 50–51)

1

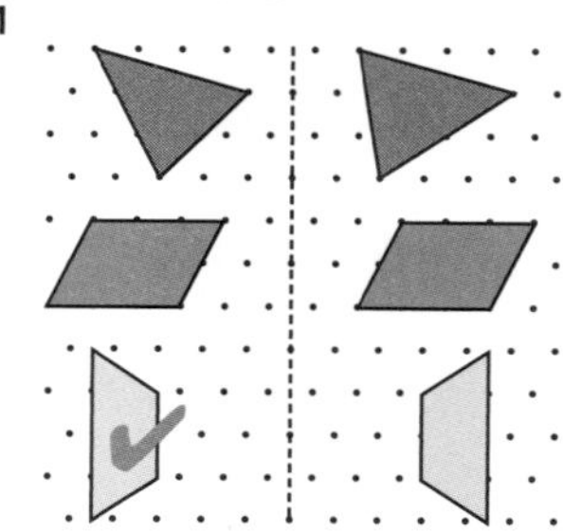

2

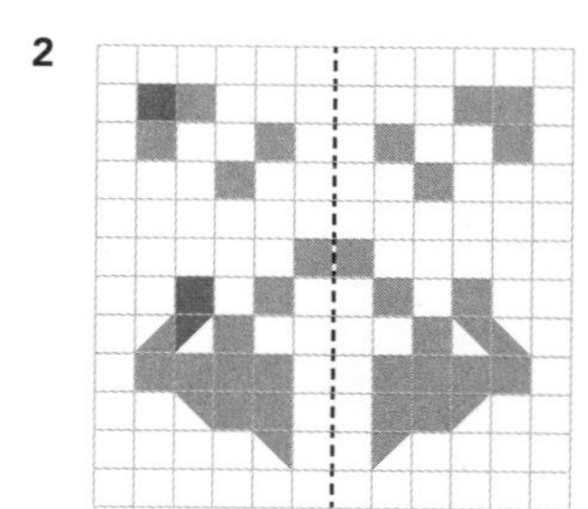

3

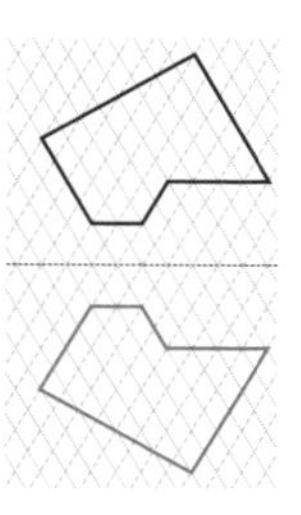

4

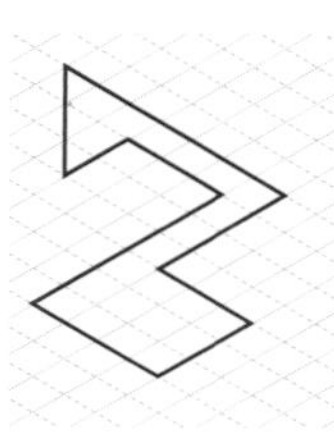

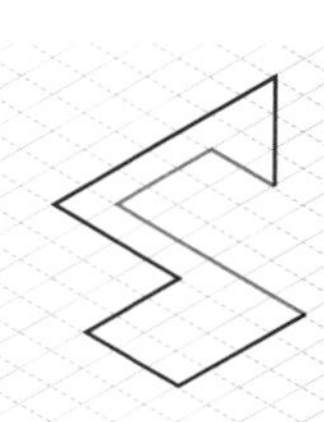

5

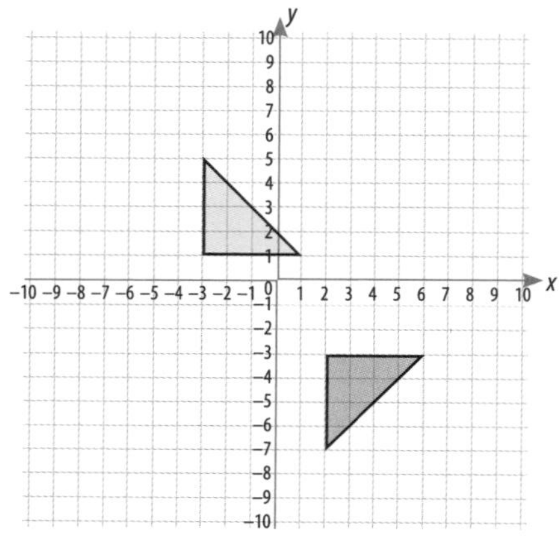

## Statistics

### Tables (page 52)

1 2nd and 3rd statements ticked

2 a)

| | Story books | Information books | Picture books | Total |
|---|---|---|---|---|
| Boys | $\frac{3}{8}$ | 48 | 25% | 128 |
| Girls | 53 or $\frac{53}{120}$ | 49 | 15% | 120 |

b) $\frac{147}{248}$

3 Award **2 marks** for all correct.

Award **1 mark** for any three correct.

| | Building repairs | Parks and outside spaces | Road repairs |
|---|---|---|---|
| 2015 | £23,300 | £12,389 | £18,273 |
| 2016 | £27,308 | £19,826 | £28,979 |
| 2017 | £19,392 | £23,785 | £36,208 |
| TOTAL | **£70,000** | **£56,000** | **£83,460** |

### Timetables (page 53)

1 Award **2 marks** for all correct.

Award **1 mark** for any three correct.

| | Stop A | Stop B | Stop C | Stop D |
|---|---|---|---|---|
| Tram 3 | 13:03 | 13:37 | 14:13 | 15:05 |

2 a) 25 minutes b) 21:38

3 Award **2 marks** for all correct.

Award **1 mark** for any two correct.

Berlin 13:50, Athens 19:05, Stockholm 15:20, Moscow 20:40

### Pictograms (page 54)

1 a) 6,750 b) $\frac{1}{3}$

2 1st and 3rd statement ticked

3 a) 1,050,000 b) 5,425,000 c) 25%

### Bar charts (page 55)

1 a)

| Peter | Remi | Amber | Jayden | Kia |
|---|---|---|---|---|
| 100% | 65% | 50% | 30% | 80% |

b)

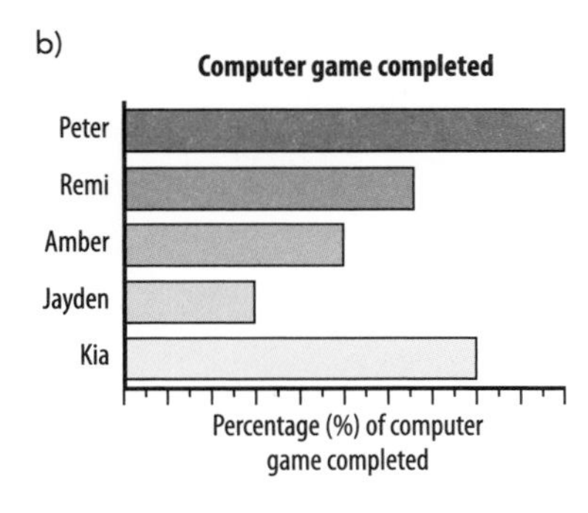


**2** Award **1 mark** for correct answer YES and an explanation either:
- the pebbles are 0.375 kg lighter than the soil and 0.375 is equivalent to $\frac{3}{8}$
- the difference between the two masses is 0.375 and this is equivalent to $\frac{3}{8}$
- the pebbles are $1\frac{1}{2}$ spaces (intervals) lower on the bar chart; one space is $\frac{1}{4}$ or $\frac{2}{8}$ kg so $1\frac{1}{2}$ spaces is $\frac{3}{8}$ kg.

**Do not award** a mark for simply restating that the pebbles are $\frac{3}{8}$ kg lighter than the soil.

**3** a) A and D  b) 2017  c) £45,000

## Pie charts (page 56)

**1** Award **2 marks** for correct answer (accept 2° margin of error). Award **1 mark** for a correctly constructed pie chart but with labels missing.

**2** a) 65  b) 780

**3** a) 1,600  b) 200  c) 300

## Line graphs (page 57)

**1** a)

| Distance (km) | 1 | 2 or 5.5 | 3.75 or 4.75 | 4 | 5.5 |
|---|---|---|---|---|---|
| Height (m) | 375 | 625 | 875 | 937.5 | 625 |

b) 2.5

**2** a) 3.75 million
b) 1.375 million (accept a value greater than 1.3 million but less than 1.4 million)
c) Accept more people are at work during the day so the number of viewers at 1 p.m. will be fewer or more people are at home in the evening and at work during the day or a similar explanation.

## Averages (page 58)

**1** 1.75
**2** 1.7
**3** 765 ml, 475 ml and 665 ml circled
**4** 1.928
**5** 7,386